Second Workshop on Natural Language Processing for Medical Conversations 2021

Online
6 June 2021

ISBN: 978-1-7138-3024-5

NAACL-HLT 2021

Natural Language Processing for Medical Conversations

The Proceedings of the Second Workshop

June 6, 2021

Introduction

Welcome to the second workshop on natural language processing for medical conversations.

Technological advancements have been transforming healthcare rapidly in the past several years. This has been further catalyzed by the COVID-19 pandemic. Several policy changes have been made by the government with added flexibility to enable remote treatment of patients. COVID-19, its symptoms, and medications are being widely discussed on social media. These discussions are also being analyzed by researchers from various perspectives. Moreover, with the availability of wearable fitness devices, these interactions are not limited to a pandemic but go much further. While medical discussions on public forums were prevalent in the past, their prevalence is now highlighted due to the scale of the pandemic.

To address healthcare consumers, Electronic Health Record (EHR) companies have been working to make health data of patients easily available to patients. More recently, technology companies are also stepping in. Healthcare providers are also making use of automatic speech recognition (ASR) and natural language understanding to understand doctor-patient conversations and generate medical documentation automatically. Finally, smart speakers are now common in households and users interact with them about personal and public health issues.

While applying NLP to open domain is getting increasingly popular, medical conversations present unique challenges and opportunities for impact. After our successful event last year, we are excited to continue the cross-pollination between NLP researchers and medical practitioners. The goal of this workshop is to discuss state-of-the-art approaches in conversational AI, as well as share insights and challenges when applied in healthcare. This is critical in order to bridge existing gaps between research and real-world product deployments, this will further shed light on future directions.

We received 19 submissions this year, and accepted 9 reviewed papers in the proceedings of the workshop. This will be a one-day workshop including keynotes, spotlight talks, posters, and panel sessions.

Organizing Committee

- Chaitanya Shivade (Amazon)

- Rashmi Gangadharaiah (Amazon)

- Spandana Gella (Amazon)

- Sandeep Konam (Abridge)

- Shaoqing Yuan (Amazon)

- Yi Zhang (Amazon)

- Parminder Bhatia (Amazon)

- Byron Wallace (Northeastern University)

Table of Contents

Conference Program

Sunday, June 6, 2021

9:00 – 9:15	Opening Remarks
9:15 – 9:50	Invited Talk 1
10:00 – 10:15	Paper Presentation: *Gathering Information and Engaging the User ComBot: A Task-Based, Serendipitous Dialog Model for Patient-Doctor Interactions* Anna Liednikova, Philippe Jolivet, Alexandre Durand-Salmon and Claire Gardent
10:15 – 10:30	Paper Presentation: *Automatic Speech-Based Checklist for Medical Simulations* Sapir Gershov, Yaniv Ringel, Erez Dvir, Tzvia Tsirilman, Elad Ben Zvi, Sandra Braun, Aeyal Raz and Shlomi Laufer
10:30 – 11:00	Break
11:00 – 11:35	Invited Talk 2
11:45 – 12:00	Paper Presentation: *Assertion Detection in Clinical Notes: Medical Language Models to the Rescue?* Betty van Aken, Ivana Trajanovska, Amy Siu, Manuel Mayrdorfer, Klemens Budde and Alexander Loeser
12:00 – 12:15	Paper Presentation: *Medically Aware GPT-3 as a Data Generator for Medical Dialogue Summarization* Bharath Chintagunta, Namit Katariya, Xavier Amatriain and Anitha Kannan
12:15 – 13:15	Lunch Break
13:15 – 13:50	Spotlight Talk Sponsor 3M
14:00 – 15:00	Poster Session
15:00 – 15:30	Break
15:30 – 16:05	Invited Talk 3
16:15 – 16:30	Best Paper Awards

Would you like to tell me more?
Generating a corpus of psychotherapy dialogues

Seyed Mahed Mousavi[1], Alessandra Cervone[2]*, Morena Danieli[1], Giuseppe Riccardi[1]

[1]Signals and Interactive Systems Lab, University of Trento, Italy
[2]Amazon Alexa AI

`{mahed.mousavi, giuseppe.riccardi}@unitn.it`

Abstract

The acquisition of a dialogue corpus is a key step in the process of training a dialogue model. In this context, corpora acquisitions have been designed either for open-domain information retrieval or slot-filling (e.g. restaurant booking) tasks. However, there has been scarce research in the problem of collecting personal conversations with users over a long period of time. In this paper we focus on the types of dialogues that are required for mental health applications. One of these types is the follow-up dialogue that a psychotherapist would initiate in reviewing the progress of a Cognitive Behavioral Therapy (CBT) intervention. The elicitation of the dialogues is achieved through textual stimuli presented to dialogue writers. We propose an automatic algorithm that generates textual stimuli from personal narratives collected during psychotherapy interventions. The automatically generated stimuli are presented as a seed to dialogue writers following principled guidelines. We analyze the linguistic quality of the collected corpus and compare the performances of psychotherapists and non-expert dialogue writers. Moreover, we report the human evaluation of a corpus-based response-selection model.

1 Introduction

The idea of developing conversational agents as Personal Healthcare Agents (PHA) (Riccardi, 2014) has gained growing attention in recent years for various domains including mental health (Fitzpatrick et al., 2017; Abd-alrazaq et al., 2019; Ali et al., 2020). Most of the conversational agents in the mental health domain are created using rule-based and simple predefined tree-based dialogue flows, resulting in limited understanding of the user input and repetitive responses by the agent. These limitations lead to shallow conversations and weak user engagement (Abd-Alrazaq et al., 2021).

The major reasons for such limitations are the complexity of conversations, the lack of dialogue data and domain knowledge. The conversations about mental state issues are very complex because they usually encompass personal feelings, user-specific situations, different spaces of entities, and emotions. In this domain, the state-of-the-art data-driven frameworks are not applicable and domain knowledge is very scarce. The two main approaches to collect dialogue data for the purpose of developing data-driven dialogue agents are either acquiring user interaction data via user simulators and hand-designed policies (Li et al., 2016), or to collect large sets of human-human conversations in different user-agnostic settings (Budzianowski et al., 2018; Gopalakrishnan et al., 2019; Zhang et al., 2018). These approaches have been used for goal-oriented agents (e.g. reservations of restaurants) or open-domain agents answering questions about a finite set of topics (e.g. news, music, weather, games etc.). However, neither of the above approaches can address the need for personal conversations which include user-specific recollections of events, objects, entities and their relations. Last but not least, state-of-the-art conversational agents cannot carry out engaging and appropriate single-user multi-session conversations. However, personal conversations' requirements include the ability of carrying out multi-session conversations over several weeks or months.

In this paper, we propose a novel methodology to collect corpora of follow-up dialogues for the mental health domain (or domains with the similar characteristics). Psychotherapists deliver interventions over a long period of time and need to monitor or react to patients' input. In this domain, dialogue follow-ups are a critical resource for psychotherapists to learn about the life events of the narrator as well as his/her corresponding thoughts and emotions in a timely manner. In Figure 1 we describe the proposed workflow for the

---

*The work was done while at the University of Trento, prior to joining Amazon Alexa AI.

Proceedings of the Second Workshop on Natural Language Processing for Medical Conversations, pages 1–9
July 6, 2021. ©2021 Association for Computational Linguistics

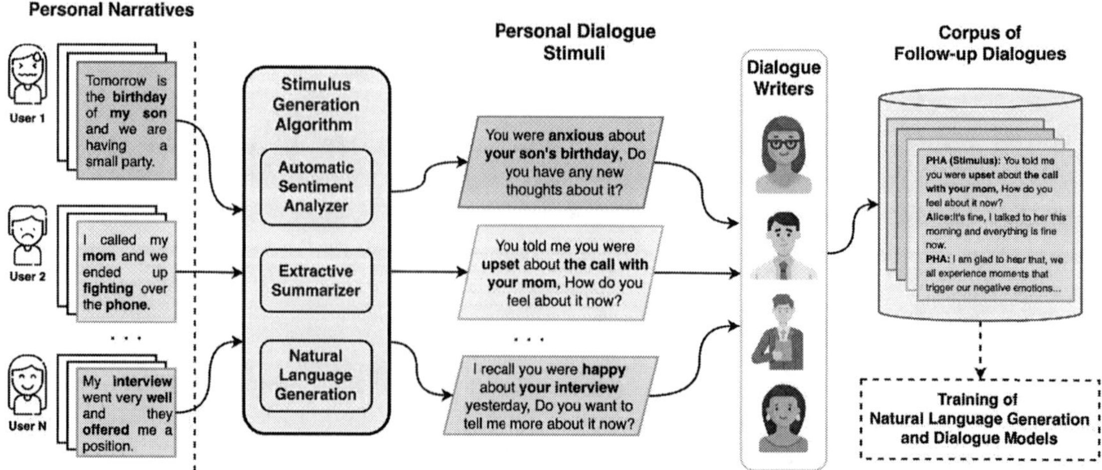

Figure 1: The workflow for the elicitation of follow-up dialogues starting from the personal narratives collected during psychotherapy (left-hand side) interventions. The stimulus generation algorithm creates a textual stimulus from personal narratives as a seed to dialogue writers. Dialogue writers use the textual stimulus and principled guidelines to generate the follow-up dialogues (right-hand side). The dialogue follow-ups may be used to train dialogue models, response-selection models and natural language generators.

acquisition of personal dialogue data aimed at training dialogue models. We first collect a dataset of personal narratives written by the users who are receiving Cognitive Behavioral Therapy (CBT) to handle their personal distress more effectively[1]. In the next step, the narratives are used to generate stimuli for the follow-up conversations with an automatic algorithm. The first part of the stimulus, the common-ground statement, contains the summary of the narrative the user has previously left and the associated emotions and the second part is a follow-up question aimed at reviewing the users life events. In the last step, the stimuli are presented to writers and they are asked to generate a conversation based on the provided stimulus by impersonating themselves as both sides of the conversation, an approach introduced firstly by Krause et al. (2017), where in our setting the sides are the PHA and the patient.

The main contributions of this paper can be summarized as follows:

- We present a methodology for data collection and elicitation of follow-up dialogues in the mental health domain.

- We present an algorithm for automatically generating conversation stimuli for follow-up dialogues in the mental health domain from a sequence of personal narratives and recollections, with a similar structure that psychotherapists use when reviewing the progress with the patient.

- We evaluate the collected dialogue corpus in terms of the quality of the obtained data, as well as the impact of domain expertise on writing the follow-up dialogues.

- We investigate the suitability of the collected corpus for developing conversational agents in the mental health domain by automatic and human evaluation of a baseline response-selection model.

2 Literature Review

Knowledge grounded dialogue corpora Previously published research have addressed the problem of collecting dialogue data starting from world knowledge facts or predefined persona descriptions. In this regard, Zhang et al. (2018) collected a dataset of conversations conditioned on synthetic persona descriptions for each side of the dialogue using Amazon Mechanical Turk (AMT) workers. Gopalakrishnan et al. (2019) collected a dataset of dialogues grounded in world knowledge by pairing AMT workers to have a conversation based on selected reading sets from Wikipedia and The Washington Post over various topics. Furthermore, Rashkin et al. (2019) have crowdsourced a dataset

---

[1]This data collection has been approved by the Ethical Committee of the University of Trento.

of conversations with implied user feelings in the context, using AMT workers where a worker writes a personal situation associated to an emotion and in the next step is paired with another worker to have a conversation about the mentioned situation. While useful for chitchat and open-domain conversations, unfortunately these resources are not a good fit to address the needs of the mental health support domain.

Mental health support dialogue corpora The research in this domain is very recent and resources are scarce. "Counseling and Psychotherapy Transcripts" published by Alexander Street Press[2] is a dataset of 4000 therapy session transcriptions on various topics, used as a resource for therapists-in-training. Pérez-Rosas et al. (2016) collected a dataset of 277 Motivational Interviewing (MI) session videos and obtained the transcriptions for each session either directly from the data source, or by recruiting AMT workers. Guntakandla and Nielsen (2018) conducted a data collection process of therapeutic dialogues in Wizard of Oz manner where the therapists were impersonating a Personal Health-care Agent. The authors recorded 324 sessions of therapeutic dialogues which were then manually transcribed. Furthermore, in the physical health coaching domain, Gupta et al. (2020) collected a dataset of conversations where the expert impersonates a PHA that engages the users into a healthier life style. For this purpose, a certified health coach interacted with 28 patients using a messaging application.

3   Dialogue Follow-Up Generation Methodology

The type of dialogues that we aim at obtaining is different from what has been reported in the literature. While previous works attempted to collect in-the-field therapeutic interactions and convert them into dialogue datasets, we present an elicitation methodology to generate a dataset of follow-up dialogues in the mental health domain, grounded in the personal narratives and with the same conversational structure that the psychotherapists use in order to review the events with the patients in a timely manner.

3.1   Collection of Personal Narratives

A group of 20 Italian native speakers who were receiving Cognitive Behavioral Therapy (CBT) were

---

[2]https://alexanderstreet.com/

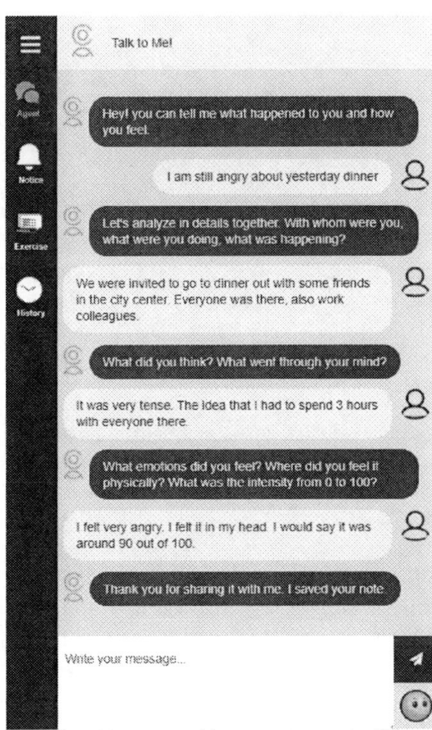

Figure 2: The user interface of the mobile application designed for collecting personal narratives (English translations). The patients were asked to describe events, persons, situations that explained their emotional arousal while answering the ABC questions designed by psychotherapists.

asked to write notes about the daily events that activated their emotional state. CBT is a psychotherapy technique based on the intuition that it is not the events that directly generate certain emotions but how these events are cognitively processed and evaluated and how irrational or dysfunctional beliefs influence this process (Oltean et al., 2017). A technique commonly used in CBT treatment is the ABC (**A**ntecedent, **B**elief, **C**onsequences). In this technique, the psychotherapist tends to identify the event that has caused the patient a certain emotion by a set of questions to define **A**) what, when and where the event happened, **B**) the patient's thoughts and beliefs about the event and **C**) the emotion the patient has experienced regarding the event. Once dysfunctional thoughts are identified, the patient is guided on how to change them or find more rational and/or functional thoughts (Sarracino et al., 2017).

We recruited 20 users who would meet with their human psychotherapists one session a week and asked them to write notes about the day-life events that caused them an emotional arousal between one

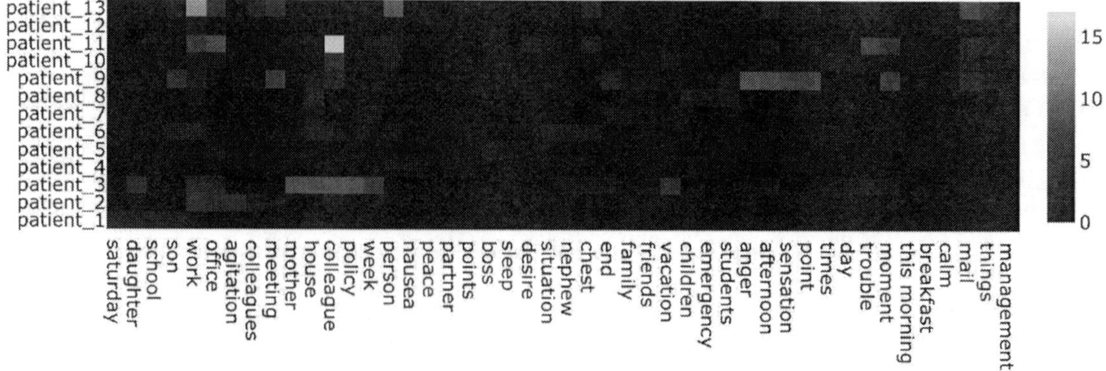

Figure 3: The heat-map of frequent nouns used by the patients in collected personal narratives (English translations). The x-axis represents the nouns extracted from the 5-most frequent list used by each user while the y-axis and z-axis represent the users and the noun frequency, respectively.

session and the following one. For this purpose, a mobile application was designed that the users could interact with for a period of three months, to answer the questions designed by the psychotherapists for the ABC technique, and assign an emotion to the note if possible. The emotions could be selected from a predefined set, equal for all users, including the six basic emotions used in psychological experiments (Happiness, Anger, Sadness, Fear, Disgust and Surprise) (Ekman, 1992), and two other complex emotional states (Embarrassment and Shame) that were considered relevant for this setting. Figure 2 shows the user interface of the application designed for this purpose.

By the end of this step, 224 ABC notes were obtained from 20 users of which 92 notes (written by 13 different subjects) are complete, i.e. the users has answered all the questions completely, and are selected for the generation of the stimuli. Considering the fact that each note, that is the answers to the ABC questions, is about a unique real-life event, we concatenate the answers in each note under the psychotherapists' supervision to convert the notes into personal narratives of one piece. Out of the 92 complete narratives, 18 narratives are assigned an emotion by the user, and 74 notes are not labeled by any emotions. A lexicon-based sentiment analyzer developed by The OpeNER project[3] is used to detect the polarity of the 74 narratives without any expressed emotions, which labeled 61 narratives as either negative or positive and 13 of them as neutral.

Lexical analysis on the selected narratives demonstrates that the language and vocabulary used in the narratives are user-specific. Figure 3 plots the recurrence of the 5 most frequent nouns used by each user in the notes, translated into English. As the figure shows, each word has been used frequently by one user and seldom by other users, indicating the personal space of entities and characteristics of the conversations in the mental health domain since the topic of these conversations, i.e. the life events and situations, varies from one patient to the other.

3.2 Generation of Personal Stimuli

We extracted one sentence from each of the 92 selected narratives using an out-of-the-shelf extractive summarizer[4], and under the supervision of the psychotherapists, designed 5 templates to convert each summary and its assigned emotion or automatically detected sentiment into a coherent stimulus consisting of a common ground and a follow-up question. For each 18 one-line narrative summaries [Summary] with an assigned emotion [Emotion] by the user, two templates are defined as;

In the notes you left previously, I read [Summary]. You told me you felt [Emotion] for that. Do you still feel [Emotion]?

I remember you told me that you felt [Emotion] because of [Summary]. How do you feel now?

while, for the 61 one-line narrative summaries with automatically determined polarity [Sentiment], two templates are defined as;

[3]https://www.opener-project.eu/

[4]sumy Automatic text summarizer, https://pypi.org/project/sumy/

Previously, you had a [Sentiment] feeling about what I read in your note [Summary]. How do you feel about it now?

I remember you had a [Sentiment] feeling about what I read in your note [Summary]. Do you have any new thoughts or considerations about it now?

and, for the 13 one-line narrative summaries without any assigned emotion or determined polarity, one template is defined as;

I read in your note about [Summary]. Do you want to tell me more about it now?

Using this methodology, we obtained 171 stimuli from the 92 selected narratives, of which 150 stimuli are used as the grounding and conversation context for follow-up dialogue generation while 21 stimuli (approximately equal to 10% of the set) are selected by stratified sampling, as a reserved subset. Table 1 shows the statistics regarding the distribution of the stimuli type used for the dialogue generation process.

3.3 Generation of Dialogue Follow-Ups

Two dialogue writer groups were recruited for the dialogue generation. The first group included 4 psychotherapists experienced in ABC therapy technique, and the second group included 4 non-expert writers. Each writer was presented with a detailed guideline including the task description as well as several examples of correct and incorrect annotation outcomes. For each provided stimulus, the writers were asked to firstly review and validate the stimulus for possible "Grammatical Error" or "Inter-sentence Incoherence" and in case of an invalid stimulus, to apply necessary modifications to correct it. Following the validation, the writers were asked to write a short dialogue follow-up based on the stimulus, assuming that the stimulus was asked by a Personal Healthcare Agent (PHA) to a user about his/her previous narrative.

The writers were asked to respect three mandatory requirements while generating the dialogues as 1) The conversation must be based on and consistent with the stimulus; 2) The flow of the conversation must be such that the user elaborates about the event introduced in the stimulus and provides more information about the event and its objects (person, location etc.) or his/her emotion to the PHA; and 3) The conversation must contain a closure turn by the

Stimulus Type	Category	Count	Total Count
with Emotion	Fear	2	32
	Happiness	9	
	Sadness	10	
	Anger	7	
	Disgust	2	
	Surprise	2	
with Valence	Positive	57	107
	Negative	50	
Neutral	-	-	11

Table 1: The distribution of the stimuli used for follow-up dialogue collection, obtained by the automatic aggregation of extracted one-line summaries, the templates and the assigned emotion or automatically detected sentiment valence.

PHA. The closure turn is an important part of the generated dialogue because these sentences play the role of the acknowledgment and grounding of the dialogue between the user and the PHA, and at the same time may increase the user willingness to use the PHA. The number of turns for the dialogues was not fixed. However, the dialogue writers were suggested to write 4 dialogue turns for each stimulus, resembling 2 turns for the user and 2 turns for the PHA (excluding the stimulus) with the last turn as the closure by the PHA. Furthermore, in order to minimize cognitive workload, the writers were suggested to distribute the work by taking a break after each 10 stimuli.

Initially, 10 stimuli were selected by stratified sampling as the Qualification Batch and were provided to all the writers for the purpose of training and resolving possible misunderstandings. The outcome of the Qualification Batch was then manually controlled and few adjustments were made with 2 of the writers. Afterwards, the rest of the stimuli were distributed such that 30% of the stimuli are annotated by all 8 writers and the rest of the stimuli are annotated by two psychotherapists and two non-expert writers.

4   Evaluation

Using the introduced elicitation methodology, we collected a corpus of follow-up conversations from the two writer groups[5]. We then performed an analysis on the obtained conversations to evaluate the

---

[5]We are currently applying for further funds to anonymize the corpus and publish a version of the corpus that respects patients' privacy and deontological requirements.

	Non-Experts	Therapists
# Dialogues	400	400
# Turns	1714	1494
# Unique Tokens	3146	4251
Avg. Turns per Dialogue	4.2	3.7

Table 2: The statistics of the collected corpus of follow-up dialogues using the proposed elicitation methodology per each writer group, non-experts and psychotherapists.

Dialogue Act	Non-Experts	Therapists
inform	1487	1777
answer	768	925
auto-positive	591	333
question	396	452
request	217	194
suggest	162	167
offer	117	26
confirm	65	36
disconfirm	56	63
address-suggest	40	17
address-request	2	9
other	77	11

Table 3: The distribution of the Dialogue Acts in the generated follow-up conversations by each writer group using ISO standard DA tagging in Italian (Roccabruna et al., 2020). Less frequent DAs to the task as accept-apology, apology, promise, accept-offer, and Feedback dimension DAs auto-negative, allo-negative and allo-positive are presented as "other" in the Table (Bunt et al., 2010).

elicitation methodology and to investigate the impact of domain expertise on the collected dialogues by comparing the performances of psychotherapists and non-expert writers.

4.1 Validation of the Generated Stimuli

In the first subtask, while 34.2% of the provided stimuli to the non-expert writers were labeled as invalid, this percentage by the psychotherapist group was 44.5%. Furthermore, the inter-annotator agreement measured by Fleiss κ coefficient (Fleiss, 1971) was higher in the latter group (0.26) as opposed to the non-expert group (0.06). This discrepancy in the validation subtask suggests that the assessment of the stimuli by each writer is affected by their level of competence in the domain and a more precise assessment of the stimuli as an effect of domain expertise. Therefore, domain expertise seems to be an important requirement for the quality of validation annotation in the mental health domain. Nevertheless, by representing each writer group by their consensus vote over the subset of stimuli for which we have a consensus decision, the inter-group agreement over this subset of 27 stimuli was 0.6639, measured by Cohen's κ coefficient (Cohen, 1960), suggesting that even though domain knowledge and expertise results in a fine-grained assessment, it is still feasible to obtain a course-grained validation over the generated stimuli with a group of non-expert writers with appropriate guidelines.

While the expert group labeled 60% of the invalid stimuli due to "Inter-sentence Incoherence" with respect to the automatic generation and combination of the stimuli elements (the summary, the sentiment, and the template), "Grammatical Error" was the assigned error in most of the stimuli labeled as invalid, 69%, by the non-expert group. Regarding the corrections applied to the invalid stimuli, modifications were mostly about the automatically extracted summary and detected polarity. The modifications on the summary sentence included refactoring the structure, re-positioning sections of the summary or restoring the punctuation. As for the modifications on the detected sentiment, while the modifications done by the non-expert writers were about changing negative and positive polarity with one another, the experts tended to be more conservative in expressing a sentiment for the stimuli as they mostly changed the stimuli with detected sentiment to neutral ones without any polarity.

In less than 10% of the cases the writers, mostly the psychotherapists, modified the template and specifically the follow-up question. In these cases, the questions were changed to a more summary-specific ones such as *"...What was the distorted thought that came to your mind?"*.

4.2 Analysis of the Dialogue Data Collection

As the result of elicitation process, we collected a dataset of follow-up dialogues in the mental health domain, presented in Table 2, consisting of 800 dialogues written by both groups. The number of turns and the number of unique tokens for each group indicate that the experts tended to write shorter conversations while they used a wider range of vocabulary in writing the conversations compared to the non-expert group. Regarding the length of the generated dialogues, in 627 conversations the writ-

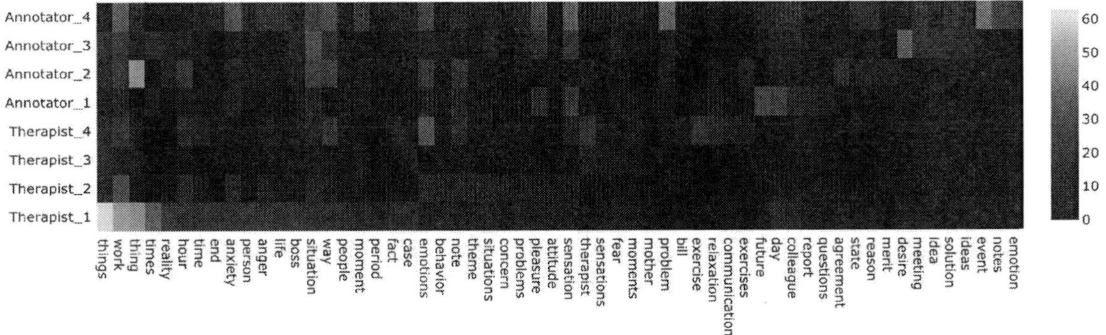

Figure 4: The heat-map of frequent nouns used by the dialogue writers in the generated conversations (English translations). The x-axis represents the nouns extracted by merging the lists of 20 most frequent nouns used per each writer. The y-axis and z-axis represent the writers and the noun frequency per each writer respectively.

ers respected the suggestion of writing 4 turns per dialogue, with exceptions of 90 dialogues written in two turns where the user replies to the stimulus and the PHA ends the conversation with a closure turn, and 83 dialogues where the user and the PHA discuss further about the event and the user's thoughts before ending the conversation.

4.2.1 Linguistic Analysis

In order to gain insights about the differences in the dialogues written be each group, we looked into the vocabulary of the nouns and entities used by each writer. Figure 4 shows the frequency heat-map of the 20 most frequent nouns used by each writer in generated dialogues, translated into English. The results indicate that the language and vocabulary used in the expert group is specific for each therapist and varies from one expert to the other, while non-expert writers have a more combined vocabulary with less inter-annotator novelty in lexicon, suggesting that the domain expertise has an influence on language and the use of vocabulary in generating conversations for the mental health domain.

Furthermore, we developed a Dialogue Act tagger to compare the conversations by their set of Dialogue Acts (DA). For this purpose, we annotated 370 of the collected dialogue follow-ups (1514 turns, approximately equal to 45% of the dataset) with the ISO standard DA tagging in Italian (Roccabruna et al., 2020) and trained an encoder–decoder model (Zhao and Kawahara, 2019) to segment each turn to its functional units and label them by their DAs. The results, presented in Table 3, show that despite the similarity in the use of the top 6 frequent DAs (inform, answer, auto-positive,

question, request and suggest), there is a diversity in the type and the frequency of the DAs used by non-expert group (such as offer, address-suggest and other less relevant DAs to the domain) with respect to the professionals, suggesting that the professionals hold a more structured conversation with respect to the other group.

4.2.2 Response-Selection Baseline

We investigated the appropriateness of the collected dialogue corpus for developing conversational agents in the mental health domain by training a TF-IDF response-selection baseline model. The model was trained on 90% of the collected conversations with a similar training setting to Lowe et al. (2015), and evaluated on the remaining 10% of the data as test set using *Recall@k* family of metrics, presented in Table 4. The model was then integrated in the application introduced in subsection 3.1 to select the correct PHA response for each user turn. 10 test users were recruited to interact with our application and write narratives about their life events by answering the ABC questions for 50 days. Each narrative was then automatically converted to a personal dialogue stimuli after one day, using the introduced methodology in subsection 3.2, to initiate a follow-up dialogue with the test user for two exchanges (4 turns) with natural language responses from the users and retrieved responses from the system. Regarding the evaluation of the dialogues, we asked the test users to assess the appropriateness and coherence of each system turn (including the stimulus) during the conversation with thumbs-up (appropriate) or thumbs-down (inappropriate) for each turn, and to evaluate the quality of the conversation as-a-whole by voting

	TF-IDF
1 in 2 R@1	0.49
1 in 10 R@1	0.21
1 in 10 R@2	0.36
1 in 10 R@5	0.55
1 in 50 R@1	0.14
1 in 50 R@2	0.18
1 in 50 R@5	0.26

Table 4: The performance of the response-selection baseline on the collected dialogue follow-ups for different recall metrics.

	Count
# Dialogues	217
# 5-star	130 (60%)
# 4-star	26 (12%)
# 3-star	41 (19%)
# 2-star	8 (3%)
# 1-star	12 (6%)
# PHA Turns	651
# Thumps-Up	594 (91%)
# Thumps-Down	57 (9%)

Table 5: The results of human evaluation of the response-selection model in follow-up dialogues. The users rated each response on a binary scale (Thumbs-Up and Thumbs-Down) as well as the whole dialogue with 1-5 star score.

from 1-star (very bad) to 5-stars (very good) for each dialogue.

The results of human evaluation on the baseline dialogue model, shown in Table 5, indicate that 91% of the system turns were considered appropriate and coherent by the test users, resulting in more than 70% of the dialogues with acceptable quality, thus suggesting the usefulness and suitability of the generated dialogues using the proposed methodology for developing PHAs in the mental health domain.

5 Conclusions

In this work, we address the need for suitable dialogue corpora to train Personal Healthcare Agents in the mental health domain. We present an elicitation methodology for dialogues in the mental health domain grounded in personal recollections. Using the proposed methodology, we collected a dataset of follow-up dialogues that psychotherapists would hold with the patients to review the personal events and emotions during a CBT intervention.

Through an analysis of the collected resource following our proposed methodology, it emerged that the task of validating responses and generating dialogues in the mental healthcare domain can be performed both by using psychotherapists and non-expert dialogue writers. Therefore, it suggests the possibility of training a larger number of non-expert dialogue writers using appropriate guidelines to obtain a valid dataset with less cost while ensuring consistency in the results.

Furthermore, we investigated the appropriateness of the collected corpus for developing conversational agents in the mental health domain. We reported automatic and human evaluation of a corpus-based response-selection baseline. We found that the test users who interacted with the model over a long-term period (50 days) considered on average 91% of system turns as appropriate and coherent, resulting into 72% of dialogues with acceptable quality.

We believe the proposed methodology can be used to tackle the problem of resource scarcity in the mental health domain. In particular, our methodology can be used to obtain corpora of dialogues grounded in personal recollections for developing dialogue models in the mental health domain.

Acknowledgements

The research leading to these results has received funding from the European Union – H2020 Programme under grant agreement 826266: COADAPT.

References

Alaa A Abd-alrazaq, Mohannad Alajlani, Ali Abdallah Alalwan, Bridgette M Bewick, Peter Gardner, and Mowafa Househ. 2019. An overview of the features of chatbots in mental health: A scoping review. *International Journal of Medical Informatics*, 132:103978.

Alaa A Abd-Alrazaq, Mohannad Alajlani, Nashva Ali, Kerstin Denecke, Bridgette M Bewick, and Mowafa Househ. 2021. Perceptions and opinions of patients about mental health chatbots: Scoping review. *Journal of Medical Internet Research*, 23(1):e17828.

Mohammad Rafayet Ali, Seyedeh Zahra Razavi, Raina Langevin, Abdullah Al Mamun, Benjamin Kane, Reza Rawassizadeh, Lenhart K. Schubert, and Ehsan Hoque. 2020. A virtual conversational agent for teens with autism spectrum disorder: Experimental results and design lessons. In *Proceedings of the 20th ACM International Conference on Intelligent*

Virtual Agents. Association for Computing Machinery.

Paweł Budzianowski, Tsung-Hsien Wen, Bo-Hsiang Tseng, Iñigo Casanueva, Stefan Ultes, Osman Ramadan, and Milica Gasic. 2018. Multiwoz-a large-scale multi-domain wizard-of-oz dataset for task-oriented dialogue modelling. In *Proceedings of the 2018 Conference on Empirical Methods in Natural Language Processing*, pages 5016–5026.

Harry Bunt, Jan Alexandersson, Jean Carletta, Jae-Woong Choe, Alex Chengyu Fang, Koiti Hasida, Kiyong Lee, Volha Petukhova, Andrei Popescu-Belis, Laurent Romary, Claudia Soria, and David Traum. 2010. Towards an iso standard for dialogue act annotation. *Seventh conference on International Language Resources and Evaluation (LREC'10)*.

Jacob Cohen. 1960. A coefficient of agreement for nominal scales. *Educational and psychological measurement*, 20(1):37–46.

Paul Ekman. 1992. Are there basic emotions? *Psychological Review*, 99(3):550–553.

Kathleen Kara Fitzpatrick, Alison Darcy, and Molly Vierhile. 2017. Delivering cognitive behavior therapy to young adults with symptoms of depression and anxiety using a fully automated conversational agent (woebot): a randomized controlled trial. *JMIR mental health*, 4(2):e19.

Joseph L Fleiss. 1971. Measuring nominal scale agreement among many raters. *Psychological bulletin*, 76(5):378.

Karthik Gopalakrishnan, Behnam Hedayatnia, Qinglang Chen, Anna Gottardi, Sanjeev Kwatra, Anu Venkatesh, Raefer Gabriel, Dilek Hakkani-Tür, and Amazon Alexa AI. 2019. Topical-chat: Towards knowledge-grounded open-domain conversations. In *INTERSPEECH*, pages 1891–1895.

Nishitha Guntakandla and Rodney Nielsen. 2018. Annotating reflections for health behavior change therapy. In *Proceedings of the Eleventh International Conference on Language Resources and Evaluation (LREC 2018)*.

Itika Gupta, Barbara Di Eugenio, Brian Ziebart, Aiswarya Baiju, Bing Liu, Ben Gerber, Lisa Sharp, Nadia Nabulsi, and Mary Smart. 2020. Human-human health coaching via text messages: Corpus, annotation, and analysis. In *Proceedings of the 21th Annual Meeting of the Special Interest Group on Discourse and Dialogue*, pages 246–256.

Ben Krause, Marco Damonte, Mihai Dobre, Daniel Duma, Joachim Fainberg, Federico Fancellu, Emmanuel Kahembwe, Jianpeng Cheng, and Bonnie Webber. 2017. Edina: Building an open domain socialbot with self-dialogues. *1st Proceedings of Alexa Prize (Alexa Prize 2017)*.

Xiujun Li, Zachary C Lipton, Bhuwan Dhingra, Lihong Li, Jianfeng Gao, and Yun-Nung Chen. 2016. A user simulator for task-completion dialogues. *arXiv preprint arXiv:1612.05688*.

Ryan Lowe, Nissan Pow, Iulian Serban, and Joelle Pineau. 2015. The Ubuntu dialogue corpus: A large dataset for research in unstructured multi-turn dialogue systems. In *Proceedings of the 16th Annual Meeting of the Special Interest Group on Discourse and Dialogue*, pages 285–294. Association for Computational Linguistics.

Horea-Radu Oltean, Philip Hyland, Frédérique Vallières, and Daniel Ovidiu David. 2017. An empirical assessment of rebt models of psychopathology and psychological health in the prediction of anxiety and depression symptoms. *Behavioural and cognitive psychotherapy*, 45(6):600–615.

Verónica Pérez-Rosas, Rada Mihalcea, Kenneth Resnicow, Satinder Singh, and Lawrence An. 2016. Building a motivational interviewing dataset. In *Proceedings of the Third Workshop on Computational Linguistics and Clinical Psychology*, pages 42–51.

Hannah Rashkin, Eric Michael Smith, Margaret Li, and Y-Lan Boureau. 2019. Towards empathetic open-domain conversation models: A new benchmark and dataset. In *Proceedings of the 57th Annual Meeting of the Association for Computational Linguistics*, pages 5370–5381. Association for Computational Linguistics.

Giuseppe Riccardi. 2014. Towards healthcare personal agents. In *Proceedings of the 2014 Workshop on Roadmapping the Future of Multimodal Interaction Research including Business Opportunities and Challenges*, pages 53–56.

Gabriel Roccabruna, Alessandra Cervone, and Giuseppe Riccardi. 2020. Multifunctional iso standard dialogue act tagging in italian. *Seventh Italian Conference on Computational Linguistics (CLiC-it)*.

Diego Sarracino, Giancarlo Dimaggio, Rawezh Ibrahim, Raffaele Popolo, Sandra Sassaroli, and Giovanni M Ruggiero. 2017. When rebt goes difficult: applying abc-def to personality disorders. *Journal of Rational Emotive & Cognitive-Behavior Therapy*, 35(3):278–295.

Saizheng Zhang, Emily Dinan, Jack Urbanek, Arthur Szlam, Douwe Kiela, and Jason Weston. 2018. Personalizing dialogue agents: I have a dog, do you have pets too? In *Proceedings of the 56th Annual Meeting of the Association for Computational Linguistics (Volume 1: Long Papers)*, pages 2204–2213. Association for Computational Linguistics.

Tianyu Zhao and Tatsuya Kawahara. 2019. Joint dialog act segmentation and recognition in human conversations using attention to dialog context. *Computer Speech & Language*, 57:108–127.

Towards Automating Medical Scribing : Clinic Visit Dialogue2Note Sentence Alignment and Snippet Summarization

Wen-wai Yim
Augmedix Inc
wenwai.yim@augmedix.com

Meliha Yetisgen
University of Washington
melihay@uw.edu

Abstract

Medical conversations from patient visits are routinely summarized into clinical notes for documentation of clinical care. The automatic creation of clinical note is particularly challenging given that it requires summarization over spoken language and multiple speaker turns; as well, clinical notes include highly technical semi-structured text. In this paper, we describe our corpus creation method and baseline systems for two NLP tasks, clinical dialogue2note sentence alignment and clinical dialogue2note snippet summarization. These two systems, as well as other models created from such a corpus, may be incorporated as parts of an overall end-to-end clinical note generation system.

1 Introduction

As a side effect of widespread electronic medical record adoption spurred by the HITECH Act, clinicians have been burdened with increased documentation demands (Tran et al.). Thus for each visit with a patient, clinicians are required to input order entries and referrals; most importantly, they are charged with the creation of a clinical note. A clinical note summarizes the discussions and plans of a medical visit and ultimately serves as a clinical communication device, as well as a record used for billing and legal purposes. To combat physician burnout, some practices employ medical scribes to assist in documentation tasks. However, hiring such assistants to audit visits and to collaborate with medical staff for electronic medical record documentation completion is costly; thus there is great interest in creating technology to automatically generate clinical notes based on clinic visit conversations.

Not only does the task of clinical note creation from medical conversation dialogue include summarizing information over multiple speakers, often the clinical note document is created with clinician-provided templates; clinical notes are also often

note	dialogue
She declines the pneumonia vaccine.	[QA-1] Doctor: Have you had a pneumonia vaccine?
	[QA-1] Patient: No, I don't think so.
	[QA-2] Doctor: Alright, do you want one?
	[QA-2] Patient: No.

Table 1: Alignment example

injected with structured information, e.g. labs. Finally, parts of clinical notes may be transcribed from dictations; or clinicians may issue commands to adjust changes in the text, e.g. "change the template", "nevermind disregard that."

In earlier work (Yim et al., 2020), we introduced a new annotation methodology that aligns clinic visit dialogue sentences to clinical note sentences with labels, thus creating sub-document granular snippet alignments between dialogue and clinical note pairs (e.g. Table 1, 2). In this paper, we extend this annotation work on a real corpus and provide the first baselines for clinic visit dialogue2note automatic sentence alignments. Much like machine translation (MT) bitext corpora alignment is instrumental to the progress in MT; we believe that dialogue2note sentence alignment will be a critical driver for AI assisted medical scribing. In the dialogue2note snippet summarization task, we provide our baselines for generating clinical note sentences from transcript snippets. Technology developed from these tasks, as well as other models generated from this annotation, can contribute as part of a larger framework that ingests automatic speech recognition (ASR) output from clinician-patient visits and generates clinical note text end-to-end (Quiroz et al., 2019).

2 Background

Table 2 depicts a full abbreviated clinical note with marked associated dialogue transcript sentences. To understand the challenges of alignment (creation of paired transcript-note input-output) and generation (creation of the note sentence from

10

Proceedings of the Second Workshop on Natural Language Processing for Medical Conversations, pages 10–20
July 6, 2021. ©2021 Association for Computational Linguistics

note	dialogue	annotations
0	Chief Complaint : 1	Evaluation of tonsil hypertrophy 2	HPI : ..	... ..	... 5	Reports enlarged tonsils, tonsil stones and sore throat. 6	Symptoms have been present for several years but have worsened over the past several months. ..	... 18	She wakes up in the morning with nausea. 19	She has frequent tonsil infections, 3-4 infections per year. ..	... ..	... 26	Physical Exam ..	... 28	Turbinates : 29	Normal size and symmetrical bilaterally ..	... ..	Tonsil : 33	3+ cryptic ..	... ..	... 62	Assessment & Plan : ..	... 68	[Risk and benefits template for tonsillectomy] ..	...	0	**Doctor:** alright enlarged tonsils. ..	... 6	**Doctor:** okay so tell me about your throat. 7	**Patient:** my tonsils they stay pretty big and they have tonsil stone and - ..	... 9	**Patient:** um like this once on this side specifically it's actually swollen- 10	**Patient:** and a couple weeks ago it was so swollen that it was like bleeding. 11	**Patient:** I wake up in the mornings and I feel like I'm going to be sick. ..	... 18	**Doctor:** so you had this for a long time? 19	**Patient:** yeah 20	**Doctor:** wait how old are you? 21	**Patient:** twenty two. 22	**Doctor:** and you've had tonsil infections since high school? 23	**Patient:** mhm. ..	... 24	**Doctor:** sore throats? 26	**Patient:** yeah. ..	... 32	**Patient:** do you think it happens more than three times in a year? 33	**Patient:** probably at least three. ..	... 48	**Doctor:** tonsils three plus cryptic . ..	... ..	... 147	**Doctor:** please insert the risks and benefits template for tonsillectomy.	note[1] → STATEMENT2SCRIBE[0] note[5] → GROUP 	STATEMENT[6], STATEMENT[7], STATEMENT[9,10]] note[6] → GROUP 	QA[18,19], QA[20,21], STATEMENT[22,23]] INCOMPLETE note[18] → STATEMENT[11] note[19] → QA[32,33] note[29] → INFFERRED-OUTSIDE note[33] → DICTATION[48] note[68] → COMMAND[147]

Table 2: Example annotations (right) for corresponding clinical note (left) and dialogue (middle). The same colors indicate matched associations.

the dialogue snippet), it is important to consider several differences in textual mediums:

Semantic variations between spoken dialogue and written clinical note narrative. Spoken language in clinic visits have vastly different representations than in highly technical clinical note reports. Dialogue may include frequent use of vernacular and verbal expressions, along with disfluencies, filler words, and false starts. In contrast, clinical note text is known to use semi-structured language, e.g. lists, and is known to have a much higher degree of nominalization. Moreover, notes frequently contain medical terminology, acronyms, and abbreviations, often with multiple word senses.

Information density and length. Whereas clinical notes are highly dense technical documents, conversation dialogue are much longer than clinical notes. In fact, in our data, dialogues were on average three times the note length. Key information in conversations are regularly interspersed.

Dialogue anaphora across multiple turns is pervasive. Anaphora is the phenomenon in which information can only be understood in conjunction with references to other expressions. Consider in the dialogue example : "Patient: I have been having swelling and pain in my knee. Doctor: How often does the knee bother you?" It's understood that the second reference of "knee" pertains to the knee-related swelling and pain. A more complex example is shown in Table 2 note line 6. While anaphora occurs in all naturally generated language, in con-versation, it may appear across multiple turns many sentences apart with contextually inferred subjects.

Order of appearance between source and target are not consistent. The order of information and organization of data in a clinical note may not match the order of discussion in a clinic visit dialogue. This provides additional challenges in the alignment process. Table 2 shows corresponding note and dialogue information with the same color.

Content incongruency. Relationship-building is a critical aspect of clinician-patient visits. Therefore visit conversations may include discussion unrelated to patient health, e.g. politics and social events. Conversely, not all clinical note content necessarily corresponds to a dialogue content. Information may come from a clinical note template or various parts of the electronic medical record.

Clinical note creation from conversation amalgamates interweaving subtasks. Elements in a clinic visit conversation (or accompanying speech introduction) are intended to be recorded or acted upon in different ways. For example, some spoken language may be directly copied to the clinical note with minor pre-determined edits, such as in a dictation, e.g. "three plus cryptic" will be converted to "3+ cryptic". However some language is meant to express directives, pertaining to adjustments to the note, e.g. "please insert the risks and benefits template for tonsillectomy." Some information is meant to be interpreted, e.g. "the pe was all normal" would allow a note sentence "CV: normal rhythm" as well as "skin: intact, no lacerations".

Finally, there are different levels of abstractive summarization over multiple statements, questions and answers as shown in the Table 2 examples.

3   Related Work

Clinical Conversation Language Understanding   Language understanding of clinical conversation can be traced to a plethora of historical work in conversation analysis regarding clinician-patient interactions (Byrne and Long, 1977; Raimbault et al., 1975; Drass, 1982; Cerny, 2007; Wang et al., 2018). More recent work has additionally included classification of dialogue utterances into semantic categories. Examples include classifying dialogue sentences into either the target SOAP section format or by using abstracted labels consistent with conversation analysis (Jeblee et al., 2019; Schloss and Konam, 2020; Wang et al., 2020). The work of (Lacson et al., 2006) framed identifying relevant parts of hemodialysis 118 nurse-patient phone conversations as an extractive summarization task. There has also been numerous works related to identifying topics, entities, attributes, and relations from clinic visit conversation – using various schemas (Jeblee et al., 2019; Rajkomar et al., 2019; Du et al., 2019). Though clinic conversation language understanding is not explored in this work, our automatic or manual sentence alignments methods produce the language understanding labels that may to used to (a) model dialogue relevance, (b) cluster dialogue topics, and (c) classify speaking mode, e.g. dictation versus question-answers.

Clinic Visit Dialogue2note Sentence Alignment   Creating a corpus of aligned clinic visit conversation dialogue sentences with corresponding clinical note sentences is instrumental for training language generation systems. Early work in this domain includes that of (Finley et al., 2018), which uses an automated algorithm based on some heuristics, e.g. string matches, and merge conditions, to align dictation parts of clinical notes. In (Yim et al., 2020), we annotated manual alignments between dialogue sentences and clinical note sentences for the entire visit; however, the dataset was small and artificial (66 visits). Here we utilize this approach on real data and additionally provide an automatic sentence alignment baseline system. To our knowledge, this is the first work to propose an automated sentence alignment system for entire clinic visit dialogue and note pairs.

Clinical Language Generation from Conversation   (Finley et al., 2018) produced dictation parts of a report, measuring performance both on gold standard transcripts and raw ASR output using statistical MT methods. In (Liu et al., 2019), the authors labeled a corpus of 101K simulated conversations and 490 nurse-patient dialogues with artificial short semi-structured summaries. They experimented with different LSTM sequence-to-sequence methods, various attention mechanisms, pointer generator mechanisms, and topic information additions. (Enarvi et al., 2020) performed similar work with sequence-to-sequence methods on a corpus of 800K orthopaedic ASR generated transcripts and notes; (Krishna et al., 2020) on a corpus of 6862 visits of transcripts annotated with clinical note summary sentences. Unlike most of previous works, our task generates clinical note sentences from labeled transcript snippets, which are at times overlapping and discontinuous. (Krishna et al., 2020)'s CLUSTER2SENT oracle system does use gold standard transcript "clusters", though different from our setup, outputs entire sections. While this strategy presupposes an upstream conversation topic segmentation system[1] as well as some extractive summarization, generation based on smaller text chunks can lead to more controllable and accurate natural language generation, critical characteristics in health applications.

4   Corpus Creation

Data   The data set was constructed from clinical encounter visits from 500 visits and 13 providers. The data for each visit consisted of a visit audio and clinical note. For each visit audio, speaker roles (e.g. clinician patient) were segmented and labeled. Automatically generated speech to text for each audio was manually corrected by annotators. Table 3 gives the summary statistics of the extracted visit audio. For all specialties, the average number of turns and sentences for transcript was 175 ± 111 and 341 ± 214, for a total of 87725 turns and 170546 sentences. The number of sentences for clinical note was 47 ± 24, for a total of 23421 sentences. Table 4 shows the number of turns and sentences per different types of speakers.

We also combined our data with external data, the mock patient visit (MPV) dataset, from (Yim

---

[1] A system that divides conversations into segments according to topics

et al., 2020) to create a total of 566 visits.[2]

specialty	providers	visits	duration	speakers
ENT	1	68	10 ± 4	4 ± 1
HAND	1	43	10 ± 4	3 ± 1
ORTHO	1	27	11 ± 5	4 ± 1
PODIATRY	4	174	7 ± 4	3 ± 1
PRIMARY	6	188	17 ± 9	4 ± 1
TOTAL	13	500	12 ± 8	4 ± 1

Table 3: Source audio statistics

Annotations Each annotation is based on a clinical note sentence association with multiple transcript sentences. A note sentence can be associated with zero transcript sentences and an INFERRED-OUTSIDE label for default template values, e.g. "cv: normal". One may also be associated with sets of transcript sentences and a set tag, e.g. DICTATION or QA (described below). Finally, when multiple sets have anaphoric references, they may be tied together using a GROUP label. Given this hierarchy, the annotation related to a single note sentence can be represented as a tree as shown in Figure 1.

Set labels
COMMAND: Spoken by the clinician to the scribe to make a change to the clinical note structure, e.g. *"add skin care macro."*
DICTATION: Spoken by the clinician to the scribe where the output text is expected to be almost verbatim, though with understood changes in abbrevations, number expressions, and language formatting commands, e.g. *"return in four to five days period."*
STATEMENT2SCRIBE: Spoken by the clinician to the scribe where information is communicated informally, e.g. *"okay so put down heart and lungs were normal"*
STATEMENT: Statements spoken by any participant in a clinic visit in natural conversation, e.g.

---

[2]To normalize for annotation differences between the Mock Patient Visits (MPV) and our corpus, we removed INFERRED-DIALOGUE labels, reattached REPEATS to a higher node, and moved all GROUP labels to the highest node.

speaker	sentences	turns
clinician_primary	99421	42480
patient	56052	36059
other	15073	9186
TOTAL	170546	87725

Table 4: Speaker statistics

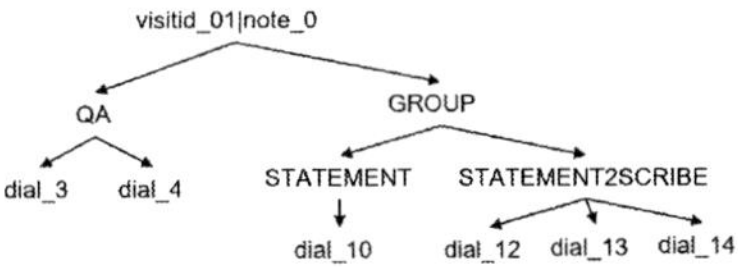

Figure 1: Annotation match tree

"it lasted about a week."
QA: Questions and answers spoken by any participant in a clinic visit in natural conversation, e.g. *"how long has the runny nose lasted? about a week."*
INFERRED-OUTSIDE: Clinical note sentences for which information comes from a known template's default value rather than the conversation, e.g. *"skin: intact."*

If after applying all possible associations and still there is information in the note sentence not available from the transcript, then an INCOMPLETE tag is added. A note sentence is left unmarked if no information can be found from the transcript. Table 2 shows label annotations with color coding for a full abbreviated transcript-note pair.

To measure interannotaor agreement, we calculated the triple, path, and span metrics introduced in (Yim et al., 2020), briefly described again here. The triple, path, and span metrics were defined based on instances constructed from the annotation tree representation. Specifically, for the triple metric, which measures unlabeled note to dialogue sentence match, instances are defined by note sentence id and transcript sentence id per visit, e.g. 'visitid_01|note_0|3'. The second metric, similar to the leaf-ancestor metric used in parsing, takes into account the full path from one note sentence to one dialogue sentence, e.g. 'visitid_01|note_0|GROUP|QA|3'. The span metric, similar to that of PARSEVAL, measures a node-level labeled span of dialogue sentences, e.g. for the top group node would be 'visitid_01|note_0|GROUP|[10,12,13,14]' (Sampson and Babarczy, 2003). When testing agreement, labels for each annotator are decomposed to these instance collections; true positive, false positive, and false negatives may be counted by the matches and mismatches between annotators. F1 score is calculated as usual. The different definitions allow both relaxed (triple) and stricter (path and span) agreement measurements.

Labeling Process A group of 11 annotators were trained for various parts of the processing task. Audio transcription was performed using Elan (archive.mpi.nl/tla/elan) and dialogue2note annotation was performed using an in-house software. Annotators underwent training on sample files for which they received in-depth feedback. They additionally took a training quiz and self-reviewed errors. After training, their interannotator agreement was calculated based on 10 final files. Their average pairwise triple, path, and span F1 scores were 0.754, 0.549, and 0.645 respectively, a reasonable performance given the task difficulty.[3]

Annotation Statistics On average 58 ± 18 % of the clinical note was marked with an annotation. This suggests that around 40% of the note is structural, e.g. blank sentences or section headers, or from outside sources, e.g. injected labs, medication lists, etc. On average 13 ± 12 % of the transcript sentences were marked. This low number suggests that much of the information from transcripts consisted of repeats or were unused. Table 5 shows that most note sentences were associated with one set type, though still many were associated with multiple. Table 6 shows the frequency of note sentences and the unique label types associated with it. From the spread of percentages for each combination category, it is apparent that understanding the entire conversation context requires combining different types of cognitive listening skills. For each note sentence, the average range of transcript sentences associated with it in the train set was 11, with the 90th percentile at 17; however there were 10% of cases with ranges above 17, which occurred when explicit topic mentions appeared far away from detailed discussion. Crossing annotations occur when content from the note and transcript appeared comparatively out of order. For example, if note sentence 0 is matched with transcript sentence 3 and meanwhile note sentence 3 is matched with transcript sentence 0, these annotations "cross", rendering automatic alignment more challenging. To quantify this, we calculate the percentages of annotations which annotates cross one, three, or five other annotations[4] (Table 7). These high percentages reveal that the order of information in the transcript differ greatly from that of the note – thus

---

# label-types	freq	%
1	8712	37
2	2914	12
3	1021	4
4	311	1
5	20	-

Table 5: Label frequency per note sentence

label-combo	note sents	%sent	% cum
{INFERRED-OUTSIDE}	3731	16	16
{STATEMENT2SCRIBE}	2664	11	27
{STATEMENT}	977	4	31
{STATEMENT2SCRIBE,INCOMPLETE}	898	4	35
{DICTATION}	742	3	38
{STATEMENT,INCOMPLETE}	706	3	41
{QA}	465	2	43
{STATEMENT,GROUP}	452	2	45
{QA,STATEMENT,GROUP}	382	2	47

Table 6: Note sentence label combination statistics

alignments are said to be non-monotonic.

The full amount of annotations from the dialogue2note labels may be used to create classifiers in many different types of tasks, e.g. dialogue relevance classification, topic segmentation, command identification, etc. However, in the remaining sections, we focus on two particular system applications : automatic dialogue2note sentence alignment and snippet summarization. For these baselines, the train and test sets were split using stratified random sampling using an 80-20 split. The training and test sets were composed of 400 and 100 of our visits; 53 and 13 for the MPV visits. 91 visits from training was reserved for development testing. As a simplification, the GROUP, INCOMPLETE, and COMMAND labels are ignored for these baselines.

crossing	percentages
cross1	33 ± 28
cross3	22 ± 27
cross5	14 ± 22

Table 7: Crossing annotation statistics

5 Sentence Alignment Baselines

We define the dialogue2note sentence alignment baseline task as the classification of 1-to-1 dialogue sentence and clinical note sentence pairs with set labels. Thus, the candidate space includes all combinations of clinical note sentences paired with all dialogue possible sentences in a visit; only those annotated with labeled associations are considered positive. This is a subset of the full annotation tasks that require 1-to-many multi-label classifications with hierarchical GROUP set labels. However, this

---

[3]These agreement values are consistent with the comparable task of simplification corpus creation, previously measured to be 0.68 kappa (Hwang et al., 2015).

[4]DICTATION, STATEMENT2SCRIBE, COMMAND labels aren't counted to focus on conversational dialogue

feature	description
match-note	Dot product of note and transcript vector divided by the magnitude of the note vector.
match-transcript	Dot product of note and transcript vector divided by the magnitude of the transcript vector.
cui-pair	UMLS concept pair, as extracted by MetaMap (Aronson and Lang, 2010), where the first concept unique identifier (cui) is from the clinical note and the second cui is from the transcript sentence. The **top_p** parameter determines which most significant cui-pair features to keep, using chi-square analysis.
prev-sent-quest	1 if the previous sentence has one of sentence has a question feature, e.g. interrogative words such as who, what etc, 0 otherwise.
jaccard-sim	If set to **local**, then defaults to jaccard similarity of the note-transcript sentence pair. If set to **regional** and similarity passes the **sim-thresh** threshold, instead, the maximum jaccard similarity from candidate regional local matches is returned. These candidate regional matches are created by by heuristically finding the closest length matches by incorporating previous and next sentences.

Table 8: Feature description for non-standard features

setup is consistent with the comparable simplification dataset creation task. We report the alignment evaluation based on pairwise F1 score. The number of positive pairwise instances in train, dev, and test sets are 19721, 4770, and 5796; including all possible negative instances 6370787, 1303972, 1706901.

Bitext Corpus Creation Related Work The topic of bitext corpus creation is often used in the context of creating resources for statistical machine translation or as a means to create cross lingual linguistic resources (Koehn, 2005; Tiedemann, 2011); it is also used to describe simplification dataset creation (Barzilay and Elhadad, 2003; Hwang et al., 2015; Štajner et al., 2017). While highly parallel bitext can be aligned using sentence length methods, much like other comparable corpora alignment strategies, multi-form comparable corpora cannot rely on monotonic ordering or correlated bitext sentence length; moreover the different text forms presents additional constraints on exact narrative structure. Like in previous work, we build our baselines for dialogue2note sentence alignment by using similarity features with some adjustment to incorporate similarity over multiple sentences.

System Description Candidate classification instances for every note sentence and transcript sentences pair were created and classified into one of the previously described set labels. For each clinical note, an additional classification instance was created for a match with an empty transcript line. (This occurs with a INFERRED-OUTSIDE label). A single tag was assigned to each classification instance according to annotated labels. If multiple tags existed per sentence pair, we took the first label in the following order: STATEMENT, STATEMENT2SCRIBE, QA, DICTATION.

Sentences were tokenized, changed to lemma form using Spacy English model (spacy.io), and vectorized according to a bag of words model. Stop words and punctuation were removed. To balance the uneven data distribution, the number of negative class instances were sampled randomly according to configurable parameter, neg_samp. We experimented with three baseline pairwise classification systems:

simple-threshold : A rule-based system that categorizes everything over threshold1 to DICTATION anything between threshold1 and threshold2 to STATEMENT2SCRIBE. These were the two labels in the train set with the highest pairwise similarities; other labels had comparable similarities.

system1 : A simple feature-based system using a decision tree classifier (scikit-learn.org). Its features included speaker category, cosine similarity, length of the note and transcript sentence vectors, and the note sentence vector. In order to take into account the match over the length of either the note or the transcript, we included a match-note and match-transcript feature described in Table 8.

system2 : A feature-based system like system1 with additional features, the transcript vector, a previous-question feature, a cui-pair feature, and a jaccard similarity feature described in Table 8. To avoid erroneous matches to answer sentences, in this system, common answers (e.g. "no") were removed from the train set.

Results After tuning, we found optimal performances for the threshold systems at threshold1=0.9 and threshold2=0.6. For system1 and system2, optimized parameters were at neg_samp=50, jaccard-sim=regional, sim-thresh=0.3, top_p=20, for a decision tree classifier. Table 9 shows the F1 results per each label. With the simple threshold system, we can see the DICTATION pairs already achieve a

label	thresh	sys1	sys2
DICTATION	0.36	0.39	0.43
STATEMENT2SCRIBE	0.20	0.36	0.36
STATEMENT	0.00	0.12	0.13
QA	0.00	0.19	0.20
INFERRED-OUTSIDE	0.00	0.59	0.66
UNMARKED	0.998	0.998	0.998

Table 9: Pairwise F1 by label

similarity	composition	thresh	sys1	sys2
0-20	0.66	0.00	0.22	0.26
20-40	0.20	0.08	0.39	0.39
40-70	0.09	0.45	0.64	0.69
70-100	0.05	0.91	0.94	0.93

Table 10: Pairwise F1 by jaccard similarity (composition is the percent of annotations within the range)

performance near that of the more complex systems. Using a simple feature based system, we see F1 measures between 0.188 and 0.390 for everything but INFERRED-OUTSIDE and UNMARKED. As expected, given the high amounts of UNMARKED, it has the highest performance. Adding additional features and curating training examples gave a minor boost across different labels as shown in the system1 and system2 differences. Analyzing the results across pairs based on similarity ranges, we see that the higher similarity pairs have higher performance, likely because the similarity features can be more reliable at those ranges (Table 10). Table 11 shows the results of system2 per label. Such results are comparable to simplification dataset creation systems with 0.33 F1 at 0-40% similarity, 0.79 F1 at 40-70%, 0.95 F1 at 70-100% (Barzilay and Elhadad, 2003).

label	gold-freq	P	R	F1
DICTATION	257	0.53	0.35	0.43
STATEMENT2SCRIBE	1248	0.32	0.43	0.36
STATEMENT	2140	0.23	0.09	0.13
QA	1239	0.25	0.16	0.20
INFERRED-OUTSIDE	912	0.72	0.61	0.66
UNMARKED	1701105	0.998	0.998	0.999

Table 11: Sys2 performance by label

Studying confusions between classes in system2, we found that overwhelmingly most errors were due to assigning unmarked passages to another label. This may be due to the simple representation of features, where certain content note or transcript bag of word features may have higher weights against similarity features. There are also cases where legitimately, the dialogue will mention what is discussed in the clinical note but is not marked in the gold standard (e.g. the same topic may be referred to multiple times but we only annotate the best instance). To a smaller extent, there were confusions among related positive class labels. Confusions between DICTATION and STATEMENT2SCRIBE occurred for high similarity sentences. Confusions between STATEMENT2SCRIBE and STATEMENT arose for cases in which dialogue may be perceived to be spoken either to a scribe or a patient, e.g. "looks normal". Confusions between STATEMENT and QA transpired because we allowed the QA label to encompass both open-ended questions, e.g. "How are you? I have been having a headache for 2 weeks" as well as very focused categorical questions, e.g. "Did you take nasal spray? No."; thus answers to open-ended questions can be easily confused with STATEMENTs.

In the current system, classifications for each note-dialogue sentence pair are labeled independently. We can improve the system by framing the required matches for each clinical note sentence as a sequence labeling problem. More semantic normalization features and surrounding sentence features would benefit the classification. Finally, in the future we can try more complex sentence vector representations.

6 Snippet Summarization Baselines

We define the snippet summarization baseline task where given the gold standard dialogue snippet text, a corresponding clinical note sentence is generated. The number of instances of aligned sets for train, dev, and test was 7129, 1851, and 2085 respectively. The average number of input and output tokens was 24 and 13 respectively.

Monolingual Text-to-Text Language Generation Related Work Monolingual monologue text-to-text language generation tasks include summarization (See et al., 2017), simplification (Štajner et al., 2017), and paraphrasing (Ma et al., 2018). The exact manner of transformation between the input and output text depends on comparative lengths, task-specific constraints, and level of abstraction.

In the area of conversational modeling, e.g. chatbots, the task is to produce appropriate dialogue responses given a prompt. In one simple classic setup, the response generation can be modeled as an information retrieval problem (Jurafsky and Martin, 2009; Ji et al., 2014). In such systems, the prompt query is processed and compared to those saved

section	BLEU				R-1				R-2				R-L			
	ret	vanilla	pg	pg-mt	ret	vanilla	pg	pg-mt	ret	vanilla	pg	pg-mt	ret	vanilla	pg	pg-mt
AP	0.26	0.19	**0.38**	**0.38**	0.20	0.14	**0.33**	0.31	0.09	0.03	0.17	**0.18**	0.19	0.13	**0.31**	0.30
CC	0.22	0.22	**0.30**	**0.30**	0.16	0.15	**0.26**	0.24	0.05	0.05	**0.11**	0.10	0.15	0.14	**0.25**	0.22
HPI	0.26	0.19	**0.36**	**0.36**	0.18	0.16	**0.32**	0.29	0.07	0.04	**0.14**	**0.14**	0.17	0.15	**0.30**	0.27
IM	0.29	0.15	0.61	**0.73**	0.42	0.17	0.65	**0.75**	0.31	0.02	0.52	**0.61**	0.41	0.16	0.63	**0.73**
PE	0.35	0.19	**0.44**	**0.44**	0.29	0.17	0.39	**0.40**	0.17	0.04	**0.24**	0.23	0.28	0.16	0.38	**0.39**
ROS	0.15	0.12	0.21	**0.22**	0.13	0.11	**0.28**	0.24	0.03	0.01	0.04	**0.06**	0.11	0.10	**0.27**	0.22
ALL	0.27	0.19	**0.38**	**0.38**	0.21	0.15	**0.33**	0.32	0.09	0.04	**0.17**	**0.17**	0.19	0.14	**0.32**	0.30

Table 12: BLEU, ROUGE-1, ROUGE-2, and ROUGE-L performance by sections

in training data. The system produces the saved response to the prompt most similar to that of the query. Although our task is not to respond a user, we may utilize the same type of system. Specifically, we can instead model the note sentence as the retrieval response to a dialogue input prompt.

Our problem most closely resembles meeting conversation summarization, in which the source data is a meeting conversation (dialogue) and the target data is a meeting summary (monologue) (Carenini et al., 2011). Method pipelines include multiple classifiers such as topic segmentation, action item identification, as well as some language generation module. There is also work with end-to-end pipelines that perform extractive and abstractive neural generation (Zhu et al., 2020; Mehdad et al., 2013). Unlike a typical summarization task, our source data is of a more comparable length, making the task more tractable. For our baselines, in addition to a simple retrieval based system, we experimented with a classic sequence-to-sequence model with and without a pointer-generator.

Note Section Identification Clinical notes are typically organized into different sections demarked by section headers as shown in Table 2 note lines 0, 2, 26, and 62. In order to report language generation performances grouped by sections and also to experiment with joint section prediction, we automatically labeled note sentences to one of six note sections using a rule-based algorithm. These categories included: History of Present Illness (HPI), Assessment and Plan (AP), Physical Exam (PE), Chief Complaint (CC), Review of Systems (ROS), and Imaging (IM). Sections headers were identified using regular expressions created by studying the train set. Subsequently, note sentences were labeled based on their corresponding section header. We modeled section prediction for two of our baseline systems : **ret, pg-mt**.

System Descriptions Below we describe our baseline systems. We trained and tested our seq-to-seq models using the LeafNATS codebase (Shi et al., 2019).

retrieval-based generator (**ret**) : Note sentence suggestion generation are modeled as a retrieval task. Paired transcript snippets and note lines (with associated section) are cached. For new transcript snippets, the note sentence corresponding to the highest cosine similarity dialogue snippet in training data is returned.

seq2seq baselines : We evaluate the performances of three sequence-to-sequence baselines with an RNN sequence encoder. The base system (**vanilla**) is a simple sequence-to-sequence system with attention. We also evaluate an option to add a pointer-generator network (**pg**). Finally, to model a pointer-generator system that outputs a summary as well as a section designation, we evaluated a final option that treats the two outputs as a multitask system (**pg-mt**).[5] Experiments were run on an EC2 p2.xlarge instance with an NVIDIA K80 GPU, taking ~150 minutes each.

Results Table 12 shows the BLEU, ROUGE-1 (R-1), ROUGE-2 (R-2), AND ROUGE-L (R-L) performances across different note sections. As shown, typically the two pointer-generator systems outperform the retrieval based and vanilla baselines. This difference may be due to the ability for the pointer-generator system to copy-and-paste items from the original input.

Comparatively, (Krishna et al., 2020)'s best CLUSTER2SENT oracle scores yielded R-1, R-2, and R-L performances of 66.5, 39.01, and 52.46, respectfully, from 6862 visits. In our low resource scenario of 566 visits, we achieved 50%, 43%, and 61% of their R-1, R-2, and R-L scores at 12% of the data. This suggests given more training data our

---

[5]Final experimental hyperparameters were set at, RNN=LSTM, batch_size=50, emb_dim=128, src_hidden_dim=256, trg_hidden_dim=256, src_seq_lens=400, trg_seq_lens=100, attn_method=luong_concat, repetition=vanilla, share_emb_weight=False.

system may similarly reach state-of-the-art levels.

Table 13 shows the accuracy of the **ret** and **pg-mt** systems for note section prediction. Although on the whole, **pg-mt** performs better than the **ret** system, for low frequency categories this is not the case. This phenomenon most likely occurs because **pg-mt** favors higher frequency labels, which is consistent with its training objective. **ret**, which classifies note section through the intermediate comparisons of input sequence similarities, is less likely to be directly skewed by class imbalances.

section	freq			acc	
	train	validation	test	ret	pg-mt
AP	1935	534	655	0.41	0.54
CC	306	71	113	0.12	0.00
HPI	3708	949	956	0.65	0.85
IM	85	7	0	0.38	0.00
PE	992	274	319	0.59	0.58
ROS	103	16	21	0.05	0.00
ALL	7129	1851	2085	0.53	0.65

Table 13: Section frequency and accuracy

Human Evaluation We sampled 10 random test snippets from each of the six section categories for evaluation (total 60 snippets). An annotator with a medical degree was asked to rank the four systems relative to each other, where 1 is the best. Additionally each system was evaluated independently with a score from 1-5 (5=best) for the categories relevancy, factual accuracy, writing-style, completeness, and overall. Table 14 shows the average scores for the different baseline systems. The vanilla seq2seq system consistently performed the worst, while the pointer-generator systems consistently performed better.

	ret	vanilla	pg	pg-mt
completeness	2.5	1.2	3.1	2.9
factual-accuracy	2.4	1.3	3.2	2.9
relevancy	2.9	1.5	3.7	3.5
writing-style	3.2	1.8	3.3	3.3
overall	2.4	1.2	3.1	2.9
rank(1=best)	2.7	3.4	1.8	2.1

Table 14: Average human evaluation ratings

While our sentence generation baselines showed modest performances, this is consistent with low resource language generation scenarios and may be ameliorated with additional training data. To improve our system, in the future, we will apply methods from low-resource machine translation techniques, utilizing unpaired sources of medical dialogue and clinic note corpora. Furthermore, we can experiment with other sequence-to-sequence approaches, e.g. transformers, for better summary generation. Joint section prediction generation may be extended to model hierarchical sections by adjusting targets to include subsections.

7   Conclusions

In this work, we provided baselines for two tasks that work towards natural language generation of note sentences from medical visit conversation. An automated dialogue2note sentence alignment system can be used to create realistic training data so immensely critical for modern systems. Meanwhile, if given properly extracted transcript snippets, dialogue2note snippet summarization could provide a valuable building block for an overall language generation system.

In future work, additional metadata information, (e.g. set labels, speaker, specialties) may be incorporated into the network architecture. Although we only explore two systems here, other models such as topic segmentation, extractive summarization, note sentence ordering, and dialogue command classification, can be trained from this annotated dataset alone. These labels may alternatively be used for additional multitask classification objectives in a full sequence-to-sequence model.

Extension of this labeled dataset may yield further interesting gains. For example, textual entailment labels between paired snippets would allow progress towards understanding and generating semantic variations and detail. Event annotation, which structures text, if performed on paired snippets, would provide training examples for data-to-text or text-to-data generation.

Together or apart, such systems would enable automation of clinical note generation whether as a full end-to-end solution or as piecemeal suggestions in a human-augmented solution. Ultimately this technology may be utilized to deburden clinicians, allowing them to focus back on patient care.

Ethical Considerations

All annotators, hired in-house, underwent HIPAA data and security training. Data was stored in dedicated HIPAA compliant compute resources. Data collection and persistence was consistent with terms of use and customer expectations. All content examples in this paper are fictitious.

References

Alan R Aronson and FranÅ§ois-Michel Lang. 2010. An overview of MetaMap: historical perspective and recent advances. *Journal of the American Medical Informatics Association : JAMIA*, 17(3):229–236.

Regina Barzilay and Noemie Elhadad. 2003. Sentence alignment for monolingual comparable corpora. In *Proceedings of the 2003 Conference on Empirical Methods in Natural Language Processing*, pages 25–32.

John F Byrne and PS Long. 1977. Doctors talking to patients. *Psychological Medicine*, 7(4):735.

Giuseppe Carenini, Gabriel Murray, and Raymond Ng. 2011. Methods for mining and summarizing text conversations. *Synthesis Lectures on Data Management*, 3(3):1–130. Publisher: Morgan & Claypool Publishers.

Miroslav Cerny. 2007. On the function of speech acts in doctor-patient communication. *Linguistica*.

Kriss A Drass. 1982. Negotiation and the structure of discourse in medical consultation. *Sociology of health& illness*.

Nan Du, Kai Chen, Anjuli Kannan, Linh Tran, Yuhui Chen, and Izhak Shafran. 2019. Extracting symptoms and their status from clinical conversations. In *Proceedings of the 57th Annual Meeting of the Association for Computational Linguistics*, pages 915–925. Association for Computational Linguistics.

Seppo Enarvi, Marilisa Amoia, Miguel Del-Agua Teba, Brian Delaney, Frank Diehl, Stefan Hahn, Kristina Harris, Liam McGrath, Yue Pan, Joel Pinto, Luca Rubini, Miguel Ruiz, Gagandeep Singh, Fabian Stemmer, Weiyi Sun, Paul Vozila, Thomas Lin, and Ranjani Ramamurthy. 2020. Generating medical reports from patient-doctor conversations using sequence-to-sequence models. In *Proceedings of the First Workshop on Natural Language Processing for Medical Conversations*, pages 22–30. Association for Computational Linguistics.

Gregory Finley, Wael Salloum, Najmeh Sadoughi, Erik Edwards, Amanda Robinson, Nico Axtmann, Michael Brenndoerfer, Mark Miller, and David Suendermann-Oeft. 2018. From dictations to clinical reports using machine translation. In *Proceedings of the 2018 Conference of the North American Chapter of the Association for Computational Linguistics: Human Language Technologies, Volume 3 (Industry Papers)*, pages 121–128, New Orleans - Louisiana. Association for Computational Linguistics.

William Hwang, Hannaneh Hajishirzi, Mari Ostendorf, and Wei Wu. 2015. Aligning sentences from standard wikipedia to simple wikipedia. In *HLT-NAACL*.

Serena Jeblee, Faiza Khan Khattak, Noah Crampton, Muhammad Mamdani, and Frank Rudzicz. 2019. Extracting relevant information from physician-patient dialogues for automated clinical note taking. In *Proceedings of the Tenth International Workshop on Health Text Mining and Information Analysis (LOUHI 2019)*, pages 65–74. Association for Computational Linguistics.

Zongcheng Ji, Z. Lu, and Hang Li. 2014. An information retrieval approach to short text conversation. *ArXiv*, abs/1408.6988.

Daniel Jurafsky and James H. Martin. 2009. *Speech and Language Processing (2nd Edition)*. Prentice-Hall, Inc., USA.

Philipp Koehn. 2005. Europarl: A parallel corpus for statistical machine translation.

K. Krishna, Sopan Khosla, Jeffrey P. Bigham, and Zachary Chase Lipton. 2020. Generating soap notes from doctor-patient conversations. *ArXiv*, abs/2005.01795.

Ronilda C. Lacson, Regina Barzilay, and William J. Long. 2006. Automatic analysis of medical dialogue in the home hemodialysis domain: Structure induction and summarization. *Journal of Biomedical Informatics*, 39(5):541–555.

Zhengyuan Liu, A. Ng, Sheldon Lee Shao Guang, AiTi Aw, and Nancy F. Chen. 2019. Topic-aware pointer-generator networks for summarizing spoken conversations. *2019 IEEE Automatic Speech Recognition and Understanding Workshop (ASRU)*, pages 814–821.

Shuming Ma, Xu Sun, Wei Li, Sujian Li, Wenjie Li, and Xuancheng Ren. 2018. Query and output: Generating words by querying distributed word representations for paraphrase generation. In *Proceedings of the 2018 Conference of the North American Chapter of the Association for Computational Linguistics: Human Language Technologies, Volume 1 (Long Papers)*, pages 196–206, New Orleans, Louisiana. Association for Computational Linguistics.

Yashar Mehdad, G. Carenini, F. Tompa, and R. Ng. 2013. Abstractive meeting summarization with entailment and fusion. In *ENLG*.

Juan C. Quiroz, Liliana Laranjo, Ahmet Baki Kocaballi, Shlomo Berkovsky, Dana Rezazadegan, and Enrico Coiera. 2019. Challenges of developing a digital scribe to reduce clinical documentation burden. *NPJ Digital Medicine*, 2.

Ginette Raimbault, Olga Cachin, Jean Marie Limal, Caroline Eliacheff, and Raphael Rappaport. 1975. Aspects of communication between patients and doctors: an analysis of the discourse in medical interviews. *Pediatrics*.

Alvin Rajkomar, Anjuli Kannan, Kai Chen, Laura Vardoulakis, Katherine Chou, Claire Cui, and Jeffrey Dean. 2019. Automatically charting symptoms from patient-physician conversations using machine learning. *JAMA Internal Medicine*, 179(6):836–838.

Geoffrey Sampson and Anna Babarczy. 2003. A test of the leaf-ancestor metric for parse accuracy | natural language engineering | cambridge core. *Natural Language Engineering*, 9(4):365–380.

Benjamin J Schloss and Sandeep Konam. 2020. Towards an automated SOAP note: Classifying utterances from medical conversations. In *Proceedings of Machine Learning Research, Machine Learning for Healthcare (MLHC)*.

A. See, Peter J. Liu, and Christopher D. Manning. 2017. Get to the point: Summarization with pointer-generator networks. In *ACL*.

Tian Shi, Ping Wang, and Chandan K Reddy. 2019. Leafnats: An open-source toolkit and live demo system for neural abstractive text summarization. In *Proceedings of the 2019 Conference of the North American Chapter of the Association for Computational Linguistics (Demonstrations)*, pages 66–71.

Jorg Tiedemann. 2011. Bitext alignment. *Synthesis Lectures on Human Language Technologies*, 4(2):1–165.

Brian D. Tran, Yunan Chen, Songzi Liu, and Kai Zheng. How does medical scribes' work inform development of speech-based clinical documentation technologies? a systematic review. 27(5):808–817.

Nan Wang, Yan Song, and Fei Xia. 2018. Constructing a Chinese medical conversation corpus annotated with conversational structures and actions. In *Proceedings of the Eleventh International Conference on Language Resources and Evaluation (LREC-2018)*, Miyazaki, Japan. European Languages Resources Association (ELRA).

Nan Wang, Yan Song, and Fei Xia. 2020. Studying challenges in medical conversation with structured annotation. In *Proceedings of the First Workshop on Natural Language Processing for Medical Conversations*, pages 12–21. Association for Computational Linguistics.

Wen-wai Yim, Meliha Yetisgen, Jenny Huang, and Micah Grossman. 2020. Alignment annotation for clinic visit dialogue to clinical note sentence language generation. In *Proceedings of The 12th Language Resources and Evaluation Conference*, pages 413–421. European Language Resources Association.

Chenguang Zhu, Ruochen Xu, Michael Zeng, and Xuedong Huang. 2020. End-to-end abstractive summarization for meetings. *ArXiv*, abs/2004.02016.

Sanja Štajner, Marc Franco-Salvador, Simone Paolo Ponzetto, Paolo Rosso, and Heiner Stuckenschmidt. 2017. Sentence alignment methods for improving text simplification systems. In *Proceedings of the 55th Annual Meeting of the Association for Computational Linguistics (Volume 2: Short Papers)*, pages 97–102. Association for Computational Linguistics.

Gathering Information and Engaging the User
ComBot : A Task-Based, Serendipitous Dialog Model
for Patient-Doctor Interactions

Anna Liednikova[&,#]**, Philippe Jolivet**[&]**, Alexandre Durand-Salmon**[&]**, Claire Gardent**[†]

[&] ALIAE
[#] Université de Lorraine
[†] CNRS/LORIA

`{philippe.jolivet,alexandre.durand-salmon}@aliae.io`
`{anna.liednikova,claire.gardent}@loria.fr`

Abstract

We focus on dialog models in the context of clinical studies where the goal is to help gather, in addition to the closed set of information collected based on a questionnaire, serendipitous information that is medically relevant. To promote user engagement and address this dual goal (collecting both a predefined set of data points and more informal information about the state of the patients), we introduce an ensemble model made of three bots: a task-based, a follow-up and a social bot. We introduce a generic method for developing follow-up bots. We compare different ensemble configurations and we show that the combination of the three bots (i) provides a better basis for collecting information than just the information seeking bot and (ii) collects information in a more efficient manner that an ensemble model combining the information seeking and the social bot.

1 Introduction

Current work on Human-Machine interaction focuses on three main types of dialogs: task-based, open domain and question answering conversational dialogs. The goal of task-based models is to gather the information needed for a given task e.g., gathering the price, location and type of a restaurant needed to recommend this restaurant. Usually trained on social media data (Roller et al., 2020) (Adiwardana et al.), open domain conversational models aim to mimick open domain conversation between two humans. Finally, question answering conversational models seek to model dialogs where a series of inter-connected questions is asked about a text passage.

In this paper, we consider dialog models in the context of clinical studies i.e., dialog models which are used to collect the information needed by the medical body to assess the impact of the clinical trial on a cohort of patients (e.g., information about their mood, their activity, their sleeping patterns). In the context of these clinical studies, the goal

of the dialog model is two-fold. A first goal is to collect a set of pre-defined data points i.e., answers to a set of pre-defined questions specified in a questionnaire. A second goal is to gather relevant serendipitous information i.e., health related information that is not addressed by the questionnaire but that is provided by the user during the interaction and which may be relevant to understand the impact of the therapy investigated by the clinical study. This requires keeping the user engaged and prompting him/her with relevant follow-up questions.

To model these three goals (collecting a predefined set of data points, keeping the user engaged and gathering more informal information about the state of the patient), we introduce an ensemble model which combines three bots: a task-based bot (MEDBOT) whose goal is to collect information about the mood, the daily life, the sleeping pattern, the anxiety level and the leisure activities of the patients; a follow-up bot (FOLLOWUPBOT) designed to extend the task-based exchanges with health-related, follow-up questions based on the user input; and an empathy bot (EMPATHYBOT) whose task is to reinforce the patient engagement by providing empathetic and socially driven feedback.

Our work makes the following contributions.

- We introduce a model where interactions are driven by three main goals: maintaining user engagement, gathering a predefined set of information units and encouraging domain related user input.

- We provide a generic method to create training data for a bot that can follow-up on the user response while remaining in a given domain (in this case the health domain).

- We show that such a follow-up bot is crucial to support both information gathering and user

Proceedings of the Second Workshop on Natural Language Processing for Medical Conversations, pages 21–29
July 6, 2021. ©2021 Association for Computational Linguistics

engagement and we provide a detailed analysis of how the three bots interact.

2 Related Work

Several approaches have explored the use of ensemble models for dialog. While Song et al. (2016) proposed an ensemble model for human-machine dialog which combines a generative and a retrieval model, further ensemble models for dialog have focused on combining agents/bots designed to model different conversation strategies. Yu et al. (2016) focus on open domain conversation and combines three agents, two to improve dialog coherence (ensuring that pronouns can be resolved and maximising semantic similarity with the current context) and one to handle topic switch (moving to a new topic when the retrieval confidence score is low). The ALANA ensemble model (Papaioannou et al., 2017b,a), developed for the Amazon Alexa Challenge i.e., for open domain chitchat, combines domain specific bots used to provide information from different sources with social bots to smooth the interactions (by asking for clarification, expressing personal views or handling profanities). Similarly, Yu et al. (2017) introduces a dialog model which interleaves a social and a task-based bot. Conversely, Gunson et al. (2020) showed that success of interleaving depends on the context and that in a public setting, users either prefer purely task-based systems or fail to see a difference between task-based and a richer ensemble model combining task-based and social bots.

Our work differs from these previous approaches in that we combine a standard, task-based model with both a social bot and a domain specific, follow-up bot. This allows both for more natural dialogs (by following up on the user input rather than systematically asking about an item in the predefined set of topics) and for additional relevant, health related information to be gathered.

3 ComBot, an ensemble Model for Repeated Task-Based Interactions

We introduce the three bots making up our ensemble model and the ensemble model combining them.

3.1 Medical Bot

MEDBOT is a retrieval model which uses the pre-trained ConveRT dialog response selection model (Henderson et al., 2019) to retrieve a query from the MedTree Corpus (Liednikova et al., 2020). It is designed to collect information from the user based on a predefined set of questions contained in a questionaire.

The MedTree Dataset. The MedTree corpus (Liednikova et al., 2020) was developed to train a task-based, information seeking, health bot on five domains: sleep, mood, anxiety, daily tasks and leisure activities. It was derived from a dialog tree provided by a domain expert (i.e., a physician) and designed to formalise typical patient-doctor interactions occurring in the context of a clinical study. In that tree, each branch captures a sequence of (Doctor Question, Patient Answer) pairs and each domain is modeled by a separate tree with the root introducing the conversation (initial question) and the leaves providing a closing statement. The MedTree corpus is then derived from this tree by extracting from each branch of the tree, all context-question pairs, where the context consists of a sequence of patient-doctor-patient turns present on that branch and the question is the following doctor question. A fragment of the decision tree created for the sleep domain and an example dialog are shown in Figure 1.

There are two versions of the MedTree corpus: one consisting of only the context/question pairs derived from the dialog tree (INIT) and the other including variants of these pairs based on paraphrases extracted from forum data (ALL). In (Liednikova et al., 2020), the ALL corpus is used to train a generative and a classification model. In our work, we use (a slightly modified version[1] of) the INIT corpus instead, as its small size facilitates retrieval (the number of candidates is small) and preliminary experimentations showed better results when using the INIT corpus.

Model. ConveRT is a Transformer-based Encoder-Decoder which is trained on Reddit (727M input-response pairs) to identify the dialog context most similar to the current context and to retrieve the dialog turn following this context. In order to retrieve from the MedTree corpus, the question that best fits the current dialog context, the MEDBOT model compares the last three turns of the current dialog with contexts from the MedTree Corpus. The model identifies the

---

[1]The modifications consists in shortening the questions, changing all leaves to statements and adding meta-statements about the dialog to account for cases where the user indicates misunderstanding or agreement

MedTree corpus context with the highest similarity score[2] and outputs the question following that context. If the selected question has already been asked in the dialog generated so far and provided it is not a question such as "What other things would you like to share with me ?", we retrieve the next best question that is not a repetition. No fine-tuning is done due to the small amount of data.

3.2 Follow-Up Bot

One main motivation behind the use of a health-bot in clinical studies is to complement the information traditionally gathered through a fixed questionnaire filled in each week by the patients with serendipitous information i.e., information that is not actively queried by the questionnaire but that is useful to analyse the cohort results.

The MEDBOT model introduced in the previous section is constrained to address only those topics which are present in the dialog tree, in effect, modeling a closed questionnaire. To allow for the collection of serendipitous health information, we develop the FOLLOWUPBOT whose function is to generate health-related questions which are not predicted by the dialog tree but which naturally follow from the user input. The main difference of FOLLOWUPBOT from MEDBOT is the way it retrieves questions that are not in the sequence, but the ones that occurs in the same context even if the question itself doesn't share the lexions with the previous turns. Rather than artificially restricting the dialog to the limited set of topics pre-defined by the dialog tree, the combined model (MEDBOT + FOLLOWUPBOT) allows for transitions based either on the dialog tree or on health-related, follow-up questions. In that sense, FOLLOWUPBOT allows not only for the collection of health-related serendipitous information but also for smoother dialog transitions.

Like MEDBOT, FOLLOWUPBOT used the pre-trained ConveRT model to retrieve context appropriate queries from a dialog dataset. In this case however, the queries are retrieved from the Health-Board dataset, a new dataset we created to support follow-up questions in the health domain.

The Healthboard Dataset. This dataset consists of (s, q) pairs where s is a (health related) state-ment and q is a follow-up question for that statement. We extract this dataset from the Health-board forum [3] as follows. We first select 16 forum categories (listed in Table 1) that are relevant to our five domains. In the forum, each category includes multiple conversational threads, each thread consists of multiple posts and each post is a text of several paragraphs that can be split into sentences. In total, we collect 175,789 posts from 31,042 threads with 5.68 posts in average per thread. We then segment each post into sentences using the default NLTK sentence segmenter. We label each sentence with a dialogue act classifier in order to distinguish statements ("sd" label) from questions ("qo" label). For this labelling, we fine-tune the Distilbert Transformer-based classification model [4] on the Switchboard Corpus Stolcke et al. (2000) using 6 classes "qo" (Open-Question), "sd" (Statement-non-opinion), "ft" (Thanking), "aa" (Agree/Accept), "%" (Uninterpretable) and "ba" (Appreciation). For each question q (i.e., sentence labelled "qo") in each thread T, we gather all statements (i.e., all sentences labeled as "sd") which precede q in T into a pool of candidate statements[5]. As dialogue turns in bots should remain short, we filter sentences that have more than 100 tokens. For each candidate statement, we calculate its similarity with the question using the dot product on their ConveRT embeddings. We filter out all candidate statements whose score with the question is less than 0.6. If after filtering the resulting pool contains at least one candidate, we select the top-ranked statement and add the statement-question pair pair to the dataset. The resulting dataset contains 3,181 (statement, question) pairs.

Model. Similar to the MEDBOT model, the FOLLOWUPBOT model used the pre-trained ConveRT model to compare the current dialog context (the preceding three turns) with the statements contained in the HealthBoard dataset using the inner product. The top-20 candidates are then retrieved and filtered using Maximal Marginal Rel-

---

[2]Both contexts are encoded using ConveRT as average of embeddings of the last turn and concatenation of preceding ones. The inner product is used to compute similarity.

[3]https://www.healthboards.com/

[4]https://huggingface.co/distilbert-base-uncased

[5]We do not restrict the set of candidates at that stage i.e., we consider all posts that precede the question within the question thread and all statements in these posts no matter how far away the statement is from the question. In practice, the set of such statements has limited size and distance does not seem to matter too much although an investigation of that factor would be interesting. We leave this question open for further research as it is not central to our paper.

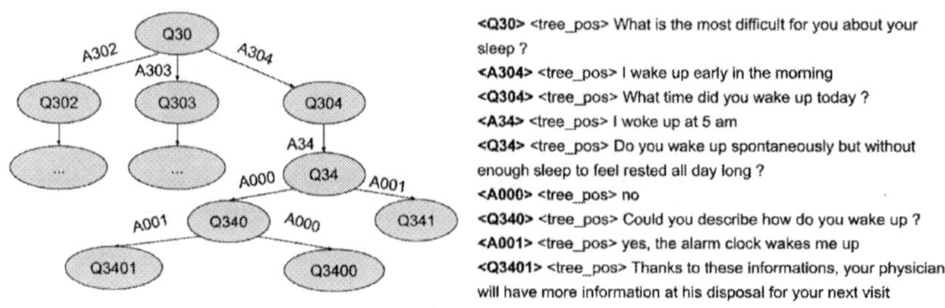

Figure 1: **Fragment of decision tree for the sleep domain and a corresponding dialog**

Category	Threads	Posts	Avg
anxiety	6852	38523	5.63
anxiety tips	42	71	1.69
chronic fatigue	670	3856	5.77
chronic pain	646	4893	7.59
depression	5327	32998	6.21
depression tips	27	51	1.89
exercise fitness	1583	8142	5.16
general health	7279	29858	4.11
healthy lifestyle	104	621	5.97
pain management	4985	38738	7.79
panic disorders	1314	8376	6.39
share your anxiety story	42	42	1
share your depression story	55	71	1.29
share your pain story	28	42	1.50
sleep disorders	1671	7656	4.59
stress	415	1973	4.76

Table 1: Forum Categories used for the Creation of the HealthBoard Dataset

evance (MMR) (Carbonell and Goldstein, 1998) with $\lambda = 0.5$ to control for repetitions[6]. Next, we compute the similarity between the remaining selected questions and the questions included in the current dialog context (all preceding dialog turns) and we exclude candidates with similarity score 0.8 or higher. After filtering, the top ranking candidate is selected and the associated follow-up question is output.

3.3 Empathy Bot

As the name suggests, the role of the EMPATHY-BOT is to engage the user by showing empathy. For this bot, we use Roller et al. (2020) generative model which was pre-trained on a variant of Reddit discussion (Baumgartner et al., 2020) and fine-tuned on the ConvAI2 (Zhang et al., 2018), Wizard of Wikipedia (Dinan et al., 2019), Empathetic Dialogues (Rashkin et al., 2019), and Blended Skill Talk datasets (BST) (Smith et al., 2020) to opti-

mize engagigness and humanness in open-domain conversation.

3.4 Ensemble Model (ComBot)

Each bot provides a single candidate. To rank them, we encode the whole current dialog context and each candidate response using the ConveRT encoder, we calculate similarity (dot product) for each candidate/context pair and we select the candidate with highest similarity score. In case all candidates scores are less than 0.1, we consider that there is no good response and we end the conversation.

4 Experiments

4.1 Data

Table 2 shows some statistics for the corpora used for pretraining (ConveRT, Blender) and for retrieval (INIT, HealthBoard). For MEDBOT and FOLLOWUPBOT, we use the ConveRT model from PolyAI [7]. For EMPATHYBOT, we use the Blender model with 90M parameters from the ParlAI library [8].

One benefit of the ensemble approach is that several models can be combined, each modelling different types of dialog requirements. We compare different configurations of our three bots: COMBOT (which combines the three bots), MEDBOT (using only the task-based bot), MED+EMPATHYBOT an ensemble model which combines the task-based (MEDBOT) and the social bot (EMPATHYBOT) and MEDBOT+ FOLLOWUPBOT, a bot combining the task-based and the follow-up question bot.

We first use automatic metrics and global satisfaction scores to compare the four models. We restrict the Acute-Eval, human-based model comparison to the two best performing systems namely,

---

[6]MMR is a measure for quantifying the extent to which a new item is both dissimilar to those already selected and similar to the target (here a selected question). A λ value of 0.5 favors similarity and diversity equally, both matter equally.

[7]https://github.com/connorbrinton/polyai-models/releases/tag/v1.0

[8]https://parl.ai/projects/recipes/

COMBOT and MEDBOT.

4.2 Evaluation

As there does not exist a dataset of well-formed
health-related dialogs whose aim is both to answer a clinical study questionaire and to allow for
serendipitous interactions, we have no test set on
which to compare the output of our dialog models.
Moreover, as has been repeatedly argued, reference-
based, automatic metrics such as BLEU or ME-
TEOR, fail to do justice to the fact that a dialog
context usually has many possible continuations.
We therefore use reference-free automatic metrics
and human assessment for evaluation.

Human evaluation.   We use the MTurk platform
to collect human-bot dialogs for our four models
(COMBOT, MEDBOT and MED+EMPATHYBOT)
and ask the crowdworkers to provide a satisfaction
rate at the end of their interaction with the bot.
We then run a second MTurk crowdsourcing task
to grade and compare dialogs pairs produced by
different models.

To collect dialogs, we ask participants to interact
with the bot for as long as they want. The conversation starts randomly with one of the initial
questions of MEDBOT. The interaction stops either when all candidates scores are less than 0.1 (cf.
Section 3.4) or when the user ends the conversation. For each model, we collect 50 dialogs. Each
annotator interacts at most once with a bot.

At the end of each human-bot conversation, the
annotator is asked to rate satisfaction on a 1-5 Lik-
ert scale (a higher score indicates more satisfaction).

Assigning a satisfaction score to a single dialog is a highly subjective task however with scores
suffering from different bias and variance per annotators (Kulikov et al., 2019). As argued by Li et al.
(2019), comparing two dialogs, each produced by
different models, and deciding on which dialog
is best with respect to a predefined set of questions, helps support a more objective evaluation.
We therefore use the Acute-Eval human evaluation
framework to compare the dialogs collected using
different bots. Since the automatic evaluation (cf.
Section 5.1) shows that COMBOT and MEDBOT
are the best systems, we compare only these two
systems asking annotators to read pairs of dialogs
created by these two bots and to then answer the
pre-defined set of questions recommended by Li
et al. (2019)'s evaluation protocol namely:

- Who would you prefer to talk to for a long
  conversation?

- If you had to say one of the speakers is interesting and one is boring, who would you say
  is more interesting?

- Which speaker sounds more human?

- Which speaker has more coherent responses
  in the conversation?

We report the percentage of time one model was
chosen over the other.

For this comparison, we consider 50 dialog pairs
(one dialog produced by COMBOT, the other by
MEDBOT) and for each Acute-Eval question, collected 50 judgments, one per dialog pair. We had
ten annotators, each annotating at most 5 dialog
pairs. To maximise similarity between the dialogs
being compared, we create the dialog pairs by computing euclidean distance between context embeddings of MEDBOT and COMBOT dialogue sets.
Then we composed a pair of two closest items and
excluded them from the choice in the next iteration.

Automatic Metrics.   After collecting dialogues
we perform their automatic evaluation. All scores
are computed on the 50 bot-human dialogs collected for a given model. Table 3 shows the result
scores averaged over 50 dialogs.

To measure *coherence*, we exploit the unsupervised model CoSim introduced by Mesgar et al.
(2019); Xu et al. (2018); Zhang et al. (2017). This
model measures the coherence of a dialog as the average of the cosine similarities between ConveRT
embedding vectors of its adjacent turns.

To assess *task success*, we count the number of
unique medical entities (Slots) mentioned. We do
this using the clinical NER-model from gthe Stanza
library (Zhang et al., 2020) [9], a model trained on
the 2010 i2b2/VA dataset (Uzuner et al., 2011) to
extract named entities denoting a medical problem,
test or treatment. We report the average number
of medical entities both per dialog and in the user
turns (to assess how much medical information
comes from the user).

Following Yu et al. (2017), we also calculate *Information gain (InfoGain)*, the average number of
unique tokens per dialog and *Conversation Length
(ConvLen)*, the average number of turns in the overall dialog.

---

[9] http://stanza.run/bio

	Reddit	ConvAI2	WoW	EmpaDial	BSD	INIT	HealthBoard
Nb of context-question pairs		211803	83011	76673	27018	168	3181
Nb of distinct turns	1.50B	267945	165213	88757	53335	154	73140
Nb of tokens	568B	3791971	2720426	2625338	912857	3688	202389
Nb of tokens per turn (Avg, Max, Min)		8.95	16.39	17.12	16.89	6.92	11.5
Vocabulary size		20707	95590	59438	52561	306	7321

Table 2: Corpus statistics (Reddit: pre-training corpus for ConveRT and the Empathy bot. ConvAI2, WoW, Empa-Dial and BSD: Datasets used to fine-tune the Empathy Bot. INIT: used for the MedBot retrieval step. HealthBoard: for FollowUp Bot Fine-Tuning and Retrieval .)

Model	Satisf.	CoSim	Slots	ConvLen	InfoGain	UserQ
MEDBOT	3.94	0.26	6.24 (1.68)	28.46	108.82 (3.82)	0.08 (4)
MEDBOT+ FOLLOWUPBOT	3.18	0.34	11.65 (3.22)	36.06	153.23 (4.25)	0.47 (23)
MEDBOT+ EMPATHYBOT	3.77	0.34	3.87 (1.46)	30.29	140.19 (4.63)	0.68 (33)
COMBOT	3.72	0.36	7.12 (2.82)	21.96	124.82 (5.68)	0.48 (24)

Table 3: Satisfaction Scores (Satisf.) and Results of the Automatic Evaluation. CoSim: Average Cosine Similarity between adjacent turns. Slots: Average Number of Medical Entities per dialogue (in brackets: average number in the user turns). ConvLen: Average Number of turns per dialog. InfoGain: Average number of unique tokens per dialog (in brackets: normalised by dialog length). UserQ: number of questions asked by Human (in bracket: total number for 50 dialogs). All metrics are averaged over the 50 Human-Bot dialogs collected for each model.

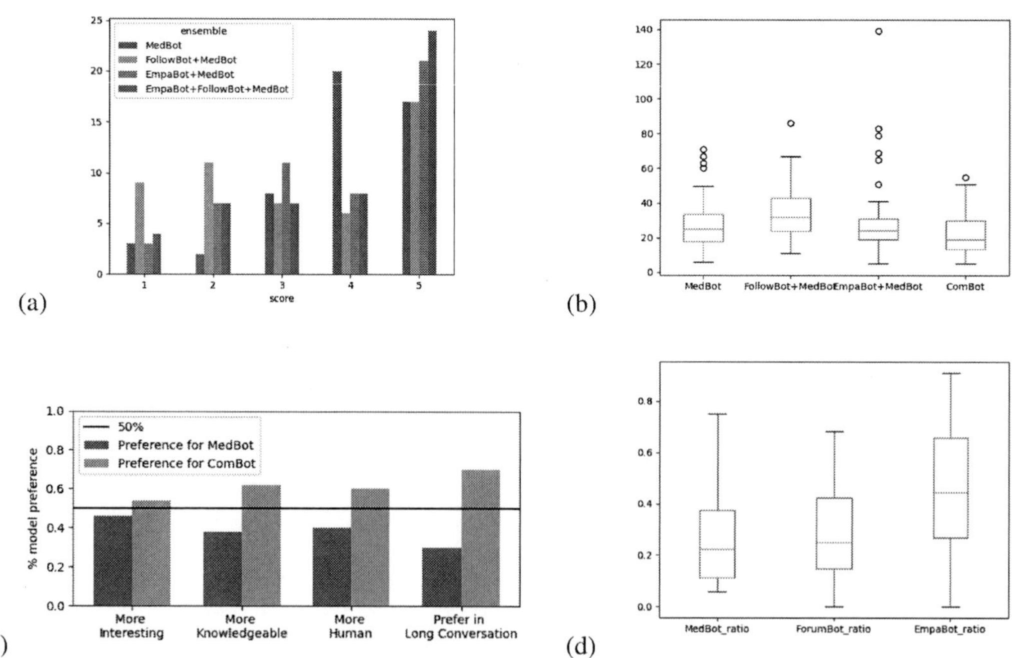

Figure 2: (a) Distribution of the Satisfaction Scores for each configuration, (b) Conversation length distribution for MedBot and ComBot, (c) Acute-Eval results for both systems, (d) Majority bot ratio in COMBOT

Finally, we compute the number of questions asked by the user (*UserQ)* as an indication of the user trust and engagement. We compute both the total number of questions present in the 50 dialog collected for a given model and the average number of question per dialog.

5   Results and Discussion

We compare four models using automatic metric and absolute satisfaction scores. Based on this first evaluation, we compare two of these models using the Acute-Eval human evaluation framework. We display an example dialog and discuss the respective use of each bot in the COMBOT model.

5.1 Automatic Evaluation and Absolute Satisfaction Scores

Table 3 shows the absolute satisfaction scores (i.e., scores provided on the basis of a single dialog rather than by comparing dialogs produced by different models) and the results of the automatic evaluation for the four models mentioned above.

ComBot provides a better basis for collecting information than MedBot. The automatic scores show that COMBOT consistently outperforms MEDBOT on informativity (Slots, InfoGrain) while allowing for shorter dialogs (ConvLen). In other words, COMBOT allows for a larger range of informational units (words and medical named entities) to be discussed in fewer turns.

ComBot collects information in a more user-friendly, more efficient manner than Med+EmpathyBot. While the InfoGain scores are higher for MED+EMPATHYBOT and MEDBOT+FOLLOWUPBOT than for COMBOT (InfoGain: 140.19 and 153.23 vs. 124.82), this is achieved at the cost of much longer dialogs (ConvLen: 30.29 and 36.06 vs. 21.96; cf. also Figure 2b) In fact, when normalising InfoGain by the number of dialog turns (ConvLen), we see that in average, a turn in COMBOT dialogs contains a much higher number of unique tokens (i.e., is more informative) than for MEDBOT (3.82), MEDBOT+EMPATHYBOT (4.63) or MEDBOT+FOLLOWUPBOT (4.25).

ComBot allows for more coherent dialogs. In terms of quality, the differences in satisfaction scores between the three models is not statistically significant ($p < 0.05$, T-test). For dialog coherence (Measured by CoSim) however, COMBOT scores highest (0.36) and the difference with MEDBOT is statistically significant ($p < 0.05$, T-test). This suggests that follow up questions help support smoother transitions between dialog turns.

5.2 Comparative Human Evaluation

The results of the comparative human evaluation are presented in Figure 2.

ComBot is judged more knowledgeable, more interesting, more human and better for long conversations. COMBOT outperforms MEDBOT on all Acute-Eval questions (Figure 2c).

In particular, users find COMBOT more knowledgeable by a large margin. This is in line with the automatic metrics results (higher COMBOT values for Slots and InfoGain) and is likely due to the fact that the COMBOT model supports the use of health-related, follow-up questions which in turn allows for a wider range of medical issues to be discussed than just those present in the MEDBOT corpus.

Users also show a clear preference for COMBOT in long conversations (Figure 2a). While this seems to contradict the fact that both models have similar satisfaction score, we conjecture that the high MEDBOT satisfaction score is an artefact of the MEDBOT model. Since the MEDBOT coverage is restricted, the users have low expectations and correspondingly give high satisfaction scores (they are easily satisfied because their expectations are low). An indication of these low user expectations is given by the number of questions asked : when users feel that the system they interact with is unrestricted, they will feel comfortable asking questions and will start to do so. Conversely, if they feel the model is restricted, they will refrain from asking questions. The results show a much higher number of questions for users interacting with COMBOT (Table 3)

5.3 Component analysis

Figure 3 displays an example Human-Bot dialog using the COMBOT model which illustrates the interactions between the three composing bots: the EMPATHYBOT closes the conversation with social chit-chat, the FOLLOWUPBOT responds to the user turn and MEDBOT asks questions from the dialog tree whenever suitable.

The proportion of turns generated by each bot (cf. Figure 2d) varies from one dialog to another, illustrating the capacity of the ensemble model to adapt to various dialog users and contexts. We find that in 55% of the collected dialogs, a majority of turns (i.e., more than 33% of the turns) is generated by the EMPATHYBOT model; in 29% of the cases by the FOLLOWUPBOT and in 16% of the cases by the MEDBOT[10]

We also observe interesting dependencies and correlations. MEDBOT is triggered twice more of-

---

[10]Since a COMBOT dialog has an average of 21 turns and only half of those are generated by the bot, this means that for 55% of the collected dialogs, the dialog contains more than 3 "social" dialog turns (turns generated by EMPATHYBOT). Similarly, 29% of the collected dialogs contain more than 3 follow-up turns (FOLLOWUPBOT) and 16% more than 3 task-based turns (MEDBOT).

BOT: : *[opening question]* What would you like to share today?
USER: : i feel soo tired
BOT: : *[FOLLOWUPBOT]* Is something making you hyper at night?
USER: : maybe i am so stressed that i cannot sleep
BOT: : *[MEDBOT]* What wakes you up at night ?
USER: : nothing, i just cannot fall asleep
BOT: : *[FOLLOWUPBOT]* Are your sleep issues anxiety related?
USER: : i think so
BOT: : *[FOLLOWUPBOT]* Also do you snore when you do sleep possibly sleep apnea?
USER: : i do not know
BOT: : *[MEDBOT]* Do you feel anxiety or stress during nights awakenings ?
USER: : yes, a bit
BOT: : *[MEDBOT]* Anxiety and sleep may be connected, it is time to speak with your physician
USER: : okay
BOT: : *[FOLLOWUPBOT]* What about getting a second opinion from another doctor about your situation?.
USER: : why not
BOT: : *[FOLLOWUPBOT]* What does your doctor say about coming off of them and then trying to go back on.
USER: : i do not know
BOT: : *[MEDBOT]* Are you not able to fall asleep regularly or punctually ?
USER: : regularly
BOT: : *[EMPATHYBOT]* i hope things get better for you , and that you are able to enjoy your time with your family .
USER: : thank you
BOT: : *[EMPATHYBOT]* you are welcome , and i wish you the best of luck with the rest of your life . :)

Figure 3: Example Human-ComBot dialog

ten after FOLLOWUPBOT (30 cases) than after EMPATHYBOT (12 cases) – this indicates that follow-up questions help bringing the user back to the questions contained in the dialog tree.

6   Conclusion

A qualitative analysis of the collected dialogs indicates several directions for further research.

Negation is often not recognised leading to interactions in which the model continues discussing a topic which was declared as irrelevant by the user. Another difficulty is knowing when to end the conversation. Long ones are good to complete the task, but bad for people who are ready to finish conversation but feel forced to continue. To improve user engagement, a possibility would be to explore whether the information provided by sentiment analysers could be exploited to help maintain a positive interaction. By detecting polarity, it could also help improve negation handling.

Another key issue concerns the emotional impact of the dialog on the user. An interaction with the bot might highlight a health issue the user was not aware of resulting in increased user stress. In such a situation, a good policy would be to provide the user with some notion of solution, some piece of information or advice which can help her face the situation and if possible, incite her to act to improve her health. Indeed some of the dialogs collected with COMBOT show that users sometimes ask for help. Here a knowledge-based agent could be useful either to provide facts that are related to the topic at hand or to highlight the connections between facts that have been mentioned in the dialog.

Acknowledgements

We thank the anonymous reviewers for their feedback. We would like to acknowledge Farnaz Ghassemi for her help in developing the FOLLOWUP-BOT. We gratefully acknowledge the support of the ALIAE company, the French National Center for Scientific Research, and the ANALGESIA Institute Foundation.

References

Daniel Adiwardana, Minh-Thang Luong, David R So, Jamie Hall, Noah Fiedel, Romal Thoppilan, Zi Yang, Apoorv Kulshreshtha, Gaurav Nemade, Yifeng Lu Quoc, and V Le.   Towards a Human-like Open-Domain Chatbot. Technical report.

Jason Baumgartner, Savvas Zannettou, Brian Keegan, Megan Squire, and Jeremy Blackburn. 2020.   The pushshift reddit dataset.

Jaime Carbonell and Jade Goldstein. 1998.   Use of MMR, diversity-based reranking for reordering documents and producing summaries.   In *SIGIR Forum (ACM Special Interest Group on Information Retrieval)*, pages 335–336, New York, New York, USA. ACM Press.

Emily Dinan, Stephen Roller, Kurt Shuster, Angela Fan, Michael Auli, and Jason Weston. 2019. Wizard of wikipedia: Knowledge-powered conversational agents.

Nancie Gunson, Weronika Sieińska, Christopher Walsh, Christian Dondrup, and Oliver Lemon. 2020. It's good to chat? evaluation and design guidelines for combining open-domain social conversation with task-based dialogue in intelligent buildings. In *Proceedings of the 20th ACM International Conference on Intelligent Virtual Agents*, IVA '20, New York, NY, USA. Association for Computing Machinery.

Matthew Henderson, Iñigo Casanueva, Nikola Mrkšić, Pei-Hao Su, Tsung-Hsien Wen, and Ivan Vulić. 2019. Convert: Efficient and accurate conversational representations from transformers.

Ilia Kulikov, Alexander H. Miller, Kyunghyun Cho, and Jason Weston. 2019. Importance of search and evaluation strategies in neural dialogue modeling.

Margaret Li, Jason Weston, and Stephen Roller. 2019. Acute-eval: Improved dialogue evaluation with optimized questions and multi-turn comparisons. *arXiv preprint arXiv:1909.03087*.

Anna Liednikova, Philippe Jolivet, Alexandre Durand-Salmon, and Claire Gardent. 2020. Learning health-bots from training data that was automatically created using paraphrase detection and expert knowledge. In *Proceedings of the International Conference on Computational Linguistics (COLING)*.

Mohsen Mesgar, Sebastian B ¨ Ucker, and Iryna Gurevych. 2019. A Neural Model for Dialogue Coherence Assessment. Technical report.

Ioannis Papaioannou, Amanda Cercas Curry, Jose Part, Igor Shalyminov, Xu Xinnuo, Yanchao Yu, Ondrej Dusek, Verena Rieser, and Oliver Lemon. 2017a. Alana: Social Dialogue using an Ensemble Model and a Ranker trained on User Feedback. In *2017 Alexa Prize Proceedings*.

Ioannis Papaioannou, Amanda Cercas Curry, Jose L Part, Igor Shalyminov, Xinnuo Xu, Yanchao Yu, Ondřej Dušek, Verena Rieser, and Oliver Lemon. 2017b. An ensemble model with ranking for social dialogue. *arXiv preprint arXiv:1712.07558*.

Hannah Rashkin, Eric Michael Smith, Margaret Li, and Y-Lan Boureau. 2019. Towards empathetic open-domain conversation models: A new benchmark and dataset. In *Proceedings of the 57th Annual Meeting of the Association for Computational Linguistics*, pages 5370–5381, Florence, Italy. Association for Computational Linguistics.

Stephen Roller, Emily Dinan, Naman Goyal, Da Ju, Mary Williamson, Yinhan Liu, Jing Xu, Myle Ott, Kurt Shuster, Eric M. Smith, Y-Lan Boureau, and Jason Weston. 2020. Recipes for building an open-domain chatbot.

Eric Michael Smith, Mary Williamson, Kurt Shuster, Jason Weston, and Y-Lan Boureau. 2020. Can you put it all together: Evaluating conversational agents' ability to blend skills.

Yiping Song, Rui Yan, Xiang Li, Dongyan Zhao, and Ming Zhang. 2016. Two are Better than One: An Ensemble of Retrieval- and Generation-Based Dialog Systems.

Andreas Stolcke, Klaus Ries, Noah Coccaro, Elizabeth Shriberg, Rebecca Bates, Daniel Jurafsky, Paul Taylor, Rachel Martin, Carol Van Ess-Dykema, and Marie Meteer. 2000. Dialogue act modeling for automatic tagging and recognition of conversational speech. *Computational Linguistics*, 26(3):339–374.

Ö. Uzuner, B.R. South, S. Shen, and S.L. DuVall. 2011. 2010 i2b2/va challenge on concepts, assertions, and relations in clinical text. *Journal of the American Medical Informatics Association*, 18(5):552–556.

Xinnuo Xu, Ondřej Dušek, Ioannis Konstas, and Verena Rieser. 2018. Better conversations by modeling, filtering, and optimizing for coherence and diversity. *Proceedings of the 2018 Conference on Empirical Methods in Natural Language Processing*.

Zhou Yu, Alan W. Black, and Alexander I. Rudnicky. 2017. Learning conversational systems that interleave task and non-task content. In *Proceedings of the 26th International Joint Conference on Artificial Intelligence*, IJCAI'17, page 4214–4220. AAAI Press.

Zhou Yu, Ziyu Xu, Alan W Black, and Alexander Rudnicky. 2016. Strategy and policy learning for non-task-oriented conversational systems. In *Proceedings of the 17th Annual Meeting of the Special Interest Group on Discourse and Dialogue*, pages 404–412, Los Angeles. Association for Computational Linguistics.

Hainan Zhang, Yanyan Lan, Jiafeng Guo, Jun Xu, and Xueqi Cheng. 2017. Reinforcing Coherence for Sequence to Sequence Model in Dialogue Generation. Technical report.

Saizheng Zhang, Emily Dinan, Jack Urbanek, Arthur Szlam, Douwe Kiela, and Jason Weston. 2018. Personalizing dialogue agents: I have a dog, do you have pets too? In *Proceedings of the 56th Annual Meeting of the Association for Computational Linguistics (Volume 1: Long Papers)*, pages 2204–2213, Melbourne, Australia. Association for Computational Linguistics.

Yuhao Zhang, Yuhui Zhang, Peng Qi, Christopher D. Manning, and Curtis P. Langlotz. 2020. Biomedical and Clinical English Model Packages in the Stanza Python NLP Library.

Automatic Speech-Based Checklist for Medical Simulations

Sapir Gershov

Technion - Israel Institute of Technology, Haifa, Israel

Dr. Yaniv Ringel, Dr. Erez Dvir, Tzvia Tsirilman, Dr. Elad Ben Zvi, Dr. Sandra Braun, Dr. Aeyal Raz

Rambam Health Care Campus, Haifa, Israel

Dr. Shlomi Laufer

Technion - Israel Institute of Technology, Haifa, Israel

`laufer@technion.ac.il`

Abstract

Medical simulators provide a controlled environment for training and assessing clinical skills. However, as an assessment platform, it requires the presence of an experienced examiner to provide performance feedback, commonly preformed using a task specific checklist. This makes the assessment process inefficient and expensive. Furthermore, this evaluation method does not provide medical practitioners the opportunity for independent training. Ideally, the process of filling the checklist should be done by a fully-aware objective system, capable of recognizing and monitoring the clinical performance. To this end, we have developed an autonomous and a fully automatic speech-based checklist system, capable of objectively identifying and validating anesthesia residents' actions in a simulation environment. Based on the analyzed results, our system is capable of recognizing most of the tasks in the checklist: F_1 score of 0.77 for all of the tasks, and F_1 score of 0.79 for the verbal tasks. Developing an audio-based system will improve the experience of a wide range of simulation platforms. Furthermore, in the future, this approach may be implemented in the operation room and emergency room. This could facilitate the development of automatic assistive technologies for these domains.

1 Introduction

In recent years, there is a growing interest in developing performance-based assessment for medical practitioners. In the pursuit for methods that may assess hands-on skills, simulation-based assessment has emerged (Srinivasan et al., 2006; Swanson et al., 1995). Simulation-based assessment requires appropriate validation metrics, and checklists are one of the most common methods. For a given simulation scenario, evaluation experts determine which actions, based on the presenting complaint, are important for the candidate to perform in order to properly manage the scenario (Scavone

et al., 2006; Morgan et al., 2007). Based on this process, a detailed checklist is developed (Morgan et al., 2007; Hilliard et al., 2000; Morgan et al., 2001; Boulet et al., 2008; Shayne et al., 2006). During the simulation, an experienced examiner is required for filling in the checklist. The need for an experienced examiner makes the assessment process very time consuming and expensive, and in addition, does not provide medical practitioners the opportunity for independent training.

Ideally, to reduce the costs of performance assessments and to allow more residents to train in a complex scenario, the process of filling the checklist should be done by a machine: A fully-aware objective system capable of recognizing and monitoring the resident performance. To this end, we have developed a end-to-end fully automatic speech-based objective checklist validation system, capable of identifying anesthesia residents' actions in a simulation environment, based solely on the participants' speech recordings. We developed a simulation setup for data collecting. The checklist system was evaluated using two different clinical scenarios for assessing skills of senior anesthesia residents. Our underlying assumption is that in many cases the communication among medical staff may represent the physical action itself. By analyzing the participants' speech, our system can automatically identify and fill the appropriate rubrics in the checklist.

2 Materials and Methods

2.1 Medical Simulation

Two clinical scenarios were developed by an experienced anesthesiologist and a medical simulation expert. The scenarios were based on scenarios previously written by the anesthesiologist (A. R) and were used for the Israeli Anesthesiology board certification exam. The first scenario included the management of a patient with a severe anaphy-

30

Proceedings of the Second Workshop on Natural Language Processing for Medical Conversations, pages 30–34
July 6, 2021. ©2021 Association for Computational Linguistics

laxis reaction and the second scenario involved a patient after surgery suffering from severe bradycardia. The study was approved by the Rambam Medical Center IRB committee.

As done in similar medical simulation studies (Hall et al., 2015; Faudeux et al., 2017; Everett et al., 2013; Wallenstein and Ander, 2015), a detailed checklist was developed for each scenario. The checklist included approximately 35 tasks the participants were expected to perform. The score for each task was in the range of 0-2, representing the performance quality in comparison to standard medical guidelines. The checklist tasks scores are scaled as follows: 0 for not observed, 1 for needs improvement, 2 for meets expectations.

Fifteen senior anesthesiology residents, 11 males and 4 females, participated in the study. Five of them preformed both simulation scenarios, 4 residents preformed only the anaphylaxis scenario and 5 preformed only the bradycardia scenario. In addition, two members of the research team played the roles of a nurse and a medical intern. During the simulation, an experienced anesthesiologist evaluated the resident's performance using the scenario checklist. A 'Laerdal' MegaCode Kelly, a full body manikin designed for the practice of Advanced Cardiovascular Life Support (ACLS), was used as the patient.

Video and audio were recorded using StreamPix digital video recording software (NorPix Inc.). The recorded video data was used by a human observer to manually fill in the checklist. For audio recordings, the resident and the nurse wore a wireless lavalier microphone transmitter (Sony UWP-D11), which was connected to a digital mixer (Tascam US-20x20). Each audio channel was saved separately.

2.2 Automatic Checklist Generation

The automatic generation of the checklist included several steps. First, automatic transcription was performed, and then, keywords were identified in each sentence. Using these keywords, a matching process between the checklist tasks and the corpus sentences was implemented. The outcome of the algorithm was a filled checklist in which the completed tasks are provided with a matching sentence and timestamp. A detailed description of each step will be provided in the following sections (Figure 2).

2.3 Automatic Transcription

The recorded audio data were automatically transcribed using Google's speech-to-Text API. This required two preprocessing steps:

1. Audio Source Separation – since the physician, nurse and intern stood in close proximity, each audio channel recorded multiple speakers as well as background noise (e.g. patient monitors). Thus, the mixed audio signal was separated into individual source signals (Vincent et al., 2018). In recent years, several open-source audio toolkits have provided implementations of audio source separation methods using deep learning (Pariente et al., 2020; Manilow et al., 2018; Ni and Mandel, 2019). In this study we used the Conv-TasNet (Luo and Mesgarani, 2019) network provided by Asteroid (Pariente et al., 2020). The network was fine-tuned for Hebrew speech as well as to our audio recording device.

2. Audio Segmentation – our objective was to provide a transcription with timestamps for each sentence. Therefore, each audio channel was segmented using the 'pyAudioAnalysis' library (Giannakopoulos, 2015). This library provides a semi-supervised audio segmentation using an SVM model. This function takes an uninterrupted audio recording as input and returns segment endpoints that correspond to the areas of "silence" between them. To achieve better division to segments, adjustments of the dynamic thresholds were performed.

2.4 Morphological and Syntactic Parsing

In order to syntactically analyze texts, the input tokens are first broken down to their constituent morphemes. However, Morphologically Rich Language (MRL), such as Hebrew, pose a unique challenge to the standard language processing pipeline. Due to extreme morphological ambiguity, global context is required in order to correctly decompose raw tokens into morphemes (More et al., 2019). To overcome this challenge, a morpho-syntactic parser for morphological and syntactic analysis of Hebrew texts (Tsarfaty et al., 2020) was used. Morphologically rich syntax parsing is useful in cases of verbs and adjectives, by reducing the variance in the transcription database.

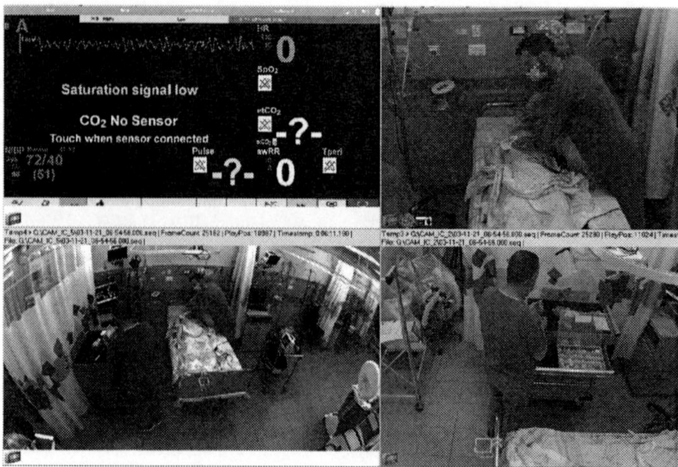

Figure 1: Data acquisition system. (A) Patient monitor; (B) Physician working area; (C) Overview of the simulation area; (D) Nurse working area. In addition, each participant carried a wireless lavalier microphone transmitter

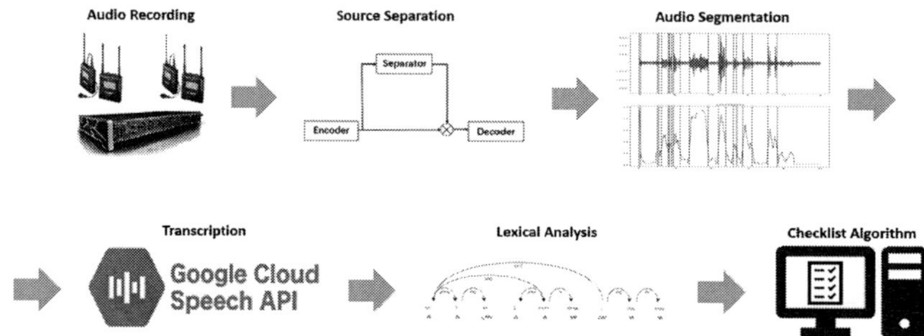

Figure 2: Automatic checklist process. End-to-end description of the checklist generation pipeline.

2.5 Word importance

The checklist includes short descriptions of each task. These descriptions guide the examiner in identifying the different assignments performed by the participants. Hence, by stripping the task description to its base form and choosing distinct words that best represent the task, a bag-of-words for each task can be generated. These "bags" will serve as touchstones for assessing how each sentence in the transcription is suitable to describe the task in hand. These keywords tend to be medical terms (medications, procedures, etc.) and combinations of an object and a verb. The matching process is based on thresholded argmax.

2.6 Checklist Evaluation

After collecting the simulation recording, a professional performance evaluator observed the video recordings and completed the checklist. As mentioned, each task in the checklist received a score in the range of 0-2. However, since the algorithm developed in this study is a binary classier, scores 1 and 2 were considered true (task preformed) and 0 was considered false. The classifier was assessed using the F_1 score (Powers, 2020).

3 Results

As mentioned in section 2.4, a proper syntactically analysis of Hebrew texts requires the disassembling of the input tokens down to their constituent morphemes. To evaluate the impact of the lexical analysis on the algorithm results, we compared two versions of the pipeline - one used the lexical analysis and the other didn't.

During the process of analyzing the data, we found that few tasks in the checklist tend to be non-verbal in their nature. Most candidate don't use any verbal commands when preforming those tasks, and the human observer can validate them only by identifying the action itself. These tasks include 'verification of intubation tube location', 'exposure of patient chest' and few others. These

Algorithm without Lexical analysis				
Division Category	Total tasks	Tasks preformed	Algorithm identified	F_1 score
All	664	405	249	0.682
Verbal	578	363	233	**0.704**
Non-Verbal	86	42	16	0.470

Table 1: Algorithm without lexical analysis performances for all, verbal and non-verbal tasks in the collected data

Algorithm with Lexical analysis				
Division Category	Total tasks	Tasks preformed	Algorithm identified	F_1 score
All	664	405	316	0.773
Verbal	578	363	292	**0.793**
Non-Verbal	86	42	24	0.585

Table 2: Algorithm with lexical analysis performances for all, verbal and non-verbal tasks in the collected data

tasks have a dramatic affect on the performance of our system. Based on this findings, we decided to divide our checklist into two different categories: verbal and non-verbal tasks. This provided us with better understanding of the system limitations. The collected data from the post-simulation human observer and the F_1 scores over all three division: All tasks, Verbal tasks and Non-verbal tasks can be found in Table 1 & 2.

4 Discussion

In this study we developed a system for automatically filling a medical simulation checklist using audio data. The system is completely autonomous and a fully automatic pipeline from the raw audio files to a complete checklist was established. The system was assessed using novel data collected for this study.

The native language of the current participants of this study is Hebrew. This poses a unique challenge common to Morphologically Rich Language. As clearly evident from the results, using lexical analysis improved our system performances, and might have a greater impact on a more complex models. We plan to expand our work to other languages in the future and assess the system performance.

The system was successful in correctly identifying most of the tasks performed by the participants. Yet, one limitation of the system is that it is currently based on keyword matching and not on a more complex model of the conversation. The method in use has limited accuracy, and in addition, only provides a binary score indication whether the task was preformed or not. For example, the current system may indicate a drug was provided but it will not assess the dosage. In order to develop a more complex algorithm, a significantly larger data base is required. We are continuously collecting data that focuses both on a larger number of participants as well as a wide range of clinical scenarios. This will expedite the development of more complex algorithms.

Developing an audio-based system has several advantages. First, it may fit to a wide range of simulation platforms including low- and high-fidelity mannequins, virtual reality, and standardized patients. Furthermore, in the future, our system could migrate from the simulation domain and be implemented in the operation room and emergency room. This could facilitate the development of automatic assistive systems for these domains.

Acknowledgements

The study was supported by the Technion's TASP-2020 grant entitled "Autonomous Medical Simulation and Training".

References

John R. Boulet, Marta Van Zanten, André De Champlain, Richard E. Hawkins, and Steven J. Peitzman. 2008. Checklist content on a standardized patient assessment: An ex post facto review. *Advances in Health Sciences Education*, 13(1):59–69.

Tobias C. Everett, Elaine Ng, Daniel Power, Christopher Marsh, Stephen Tolchard, Anna Shadrina, and Matthew D. Bould. 2013. The Managing Emergencies in Paediatric Anaesthesia global rating scale is a reliable tool for simulation-based assessment in pediatric anesthesia crisis management.

Camille Faudeux, Antoine Tran, Audrey Dupont, Jonathan Desmontils, Isabelle Montaudié, Jean Bréaud, Marc Braun, Jean Paul Fournier, Etienne Bérard, Noémie Berlengi, Cyril Schweitzer, Hervé Haas, Hervé Caci, Amélie Gatin, and Lisa Giovannini-Chami. 2017. Development of Reliable and Validated Tools to Evaluate Technical Resuscitation Skills in a Pediatric Simulation Setting: Resuscitation and Emergency Simulation Checklist for Assessment in Pediatrics. *Journal of Pediatrics*, 188:252–257.

Theodoros Giannakopoulos. 2015. PyAudioAnalysis: An open-source python library for audio signal analysis. *PLoS ONE*, 10(12):1–17.

Andrew Koch Hall, Jeffrey Damon Dagnone, Lauren Lacroix, William Pickett, and Don Albert Klinger. 2015. Queen's simulation assessment tool: Development and validation of an assessment tool for resuscitation objective structured clinical examination stations in emergency medicine.

Robert I. Hilliard, Susan E. Tallett, and Diana Tabak. 2000. Use of an Objective Structured Clinical Examination as a Certifying Examination in pediatrics.

Yi Luo and Nima Mesgarani. 2019. Conv-TasNet: Surpassing Ideal Time-Frequency Magnitude Masking for Speech Separation. *IEEE/ACM Transactions on Audio Speech and Language Processing*, 27(8):1256–1266.

Ethan Manilow, Prem Seetharaman, and Bryan Pardo. 2018. The northwestern university source separation library. *Proceedings of the 19th International Society for Music Information Retrieval Conference, ISMIR 2018*, pages 297–305.

Amir More, Amit Seker, Victoria Basmova, and Reut Tsarfaty. 2019. Joint Transition-Based Models for Morpho-Syntactic Parsing: Parsing Strategies for MRLs and a Case Study from Modern Hebrew. *Transactions of the Association for Computational Linguistics*, 7(2001):33–48.

Pamela J. Morgan, Doreen Cleave-Hogg, and Cameron B. Guest. 2001. A comparison of global ratings and checklist scores from an undergraduate assessment using an anesthesia simulator. *Academic Medicine*, 76(10):1053–1055.

Pamela J. Morgan, Jenny Lam-McCulloch, Jodi Herold-McIlroy, and Jordan Tarshis. 2007. Simulation performance checklist generation using the Delphi technique. *Canadian Journal of Anesthesia*, 54(12):992–997.

Zhaoheng Ni and Michael I. Mandel. 2019. Onssen: an Open-Source Speech Separation and Enhancement Library. *arXiv*.

Manuel Pariente, Samuele Cornell, Joris Cosentino, Sunit Sivasankaran, Efthymios Tzinis, Jens Heitkaemper, Michel Olvera, Fabian Robert Stöter, Mathieu Hu, Juan M. Martín-Doñas, David Ditter, Ariel Frank, Antoine Deleforge, and Emmanuel Vincent. 2020. Asteroid: The PyTorch-based audio source separation toolkit for researchers. *Proceedings of the Annual Conference of the International Speech Communication Association, INTERSPEECH*, 2020-Octob(1):2637–2641.

David M. W. Powers. 2020. Evaluation: from precision, recall and F-measure to ROC, informedness, markedness and correlation. pages 37–63.

Barbara M. Scavone, Michele T. Sproviero, Robert J. McCarthy, Cynthia A. Wong, John T. Sullivan, Viva J. Siddall, and Leonard D. Wade. 2006. Development of an objective scoring system for measurement of resident performance on the human patient simulator. *Anesthesiology*, 105(2):260–266.

Philip Shayne, Fiona Gallahue, Stephan Rinnert, Craig L. Anderson, Gene Hern, and Eric Katz. 2006. Reliability of a Core Competency Checklist Assessment in the Emergency Department: The Standardized Direct Observation Assessment Tool. *Academic Emergency Medicine*, 13(7):727–732.

Malathi Srinivasan, Judith C. Hwang, Daniel West, and Peter M. Yellowlees. 2006. Assessment of clinical skills using simulator technologies.

David B. Swanson, Geoffrey R. Norman, and Robert L. Linn. 1995. Performance-Based Assessment: Lessons From the Health Professions. *Educational Researcher*, 24(5):5–11.

Reut Tsarfaty, Amit Seker, Shoval Sadde, and Stav Klein. 2020. *What's wrong with Hebrew nlp? And how to make it right.*

Emmanuel Vincent, Tuomas Virtanen, and Sharon Gannot, editors. 2018. *Audio Source Separation and Speech Enhancement.* John Wiley & Sons Ltd, Chichester, UK.

Joshua Wallenstein and Douglas Ander. 2015. Objective structured clinical examinations provide valid clinical skills assessment in emergency medicine education.

Assertion Detection in Clinical Notes:
Medical Language Models to the Rescue?

Betty van Aken[1], Ivana Trajanovska[1], Amy Siu[1],
Manuel Mayrdorfer[2], Klemens Budde[2] and Alexander Löser[1]

[1]Beuth University of Applied Sciences Berlin, [2]Charité Universitätsmedizin Berlin

ivtrajanovska@gmail.com
{bvanaken, siu, aloeser}@beuth-hochschule.de
{manuel.mayrdorfer, klemens.budde}@charite.de

Abstract

In order to provide high-quality care, health professionals must efficiently identify the presence, possibility, or absence of symptoms, treatments and other relevant entities in free-text clinical notes. Such is the task of assertion detection – to identify the assertion class (*present*, *possible*, *absent*) of an entity based on textual cues in unstructured text. We evaluate state-of-the-art medical language models on the task and show that they outperform the baselines in all three classes. As transferability is especially important in the medical domain we further study how the best performing model behaves on unseen data from two other medical datasets. For this purpose we introduce a newly annotated set of 5,000 assertions for the publicly available MIMIC-III dataset. We conclude with an error analysis that reveals situations in which the models still go wrong and points towards future research directions.

1 Introduction

The clinical information buried in narrative reports is difficult for humans to access for clinical, teaching, or research purposes (Perera et al., 2013). To provide high-quality patient care, health professionals need to have better and faster access to crucial information in a summarized and interpretable format. In this paper, we focus on English discharge summaries and the task of assertion detection, which is the classification of clinical information as demonstrated in Figure 1.

Given a piece of text, we need to identify two pieces of information – a medical entity and textual cues indicating the presence or absence of that entity. Medical entity extraction has been studied extensively (Lewis et al., 2020), we thus focus our work on the task of predicting the *present / possible / absent* class over a medical entity, addressing an important information need of

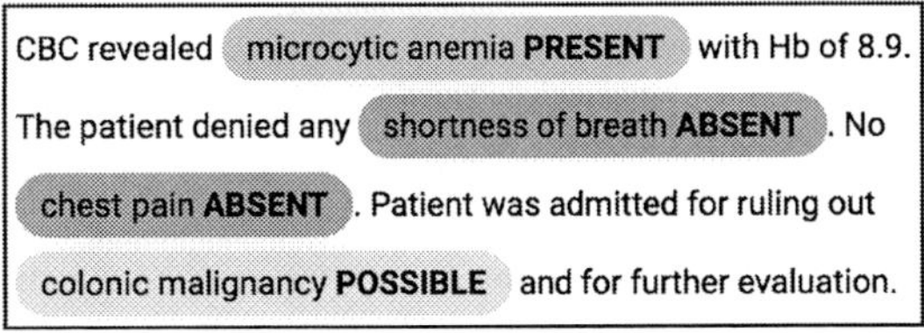

Figure 1: Sample output of our demo system. Detected entities are highlighted in red, yellow, and green to indicate *present*, *possible*, and *absent*.

health professionals. This setting is reflected in the dataset released by the 2010 i2b2 Challenge Assertions Task (de Bruijn et al., 2011a), on which we base our main evaluation.

Clinical assertion detection is known to be a difficult task (Chen, 2019) due to the free-text format of considered clinical notes. Detecting *possible* assertions is particularly challenging, because they are often vaguely expressed, and they occur far less frequently than *present* and *absent* assertions. Language models pre-trained on medical data have shown to create useful representations for a multitude of tasks in the domain (Peng et al., 2019). We apply them to our setup of assertion detection to evaluate whether they can increase performance (especially on the minority class) and where they still need improvement.

We argue that clinical assertion detection models must be transferable to data that differs from the training data, e.g. due to different writing styles of health professionals from other clinics or from other medical fields. As existing datasets do not represent such diversity, we manually annotate 5,000 assertions in clinical notes from several fields in the publicly available MIMIC-III dataset. We then use these annotated notes as an additional evaluation set to test the transferability of the best performing model.

Proceedings of the Second Workshop on Natural Language Processing for Medical Conversations, pages 35–40
July 6, 2021. ©2021 Association for Computational Linguistics

		present	*possible*	*absent*
2010 i2b2 Challenge Assertion Task	discharge summaries	21,064	1,418	6,144
BioScope	scientific publications	–	3,474	2,161
MIMIC-III Clinical Database (New)	discharge summaries	2,610	250	980
	physician letters	204	34	66
	nurse letters	293	14	59
	radiology reports	249	40	130

Table 1: Distribution of text types and classes in the three employed datasets. Note that *possible* is a minority class across datasets as well as text types. In the i2b2 dataset, for instance, only 5% of all labels are *possible*.

Our **contributions** are summarized as follows:
1) We evaluate medical language models on assertion detection in clinical notes and show that they clearly outperform previous baselines. We further study the transferability of such models to clinical text from other medical areas.
2) We manually annotate 5,000 assertions for the MIMIC-III Clinical Database (Johnson et al., 2016). We release the annotations to the research community[1] to tackle the problem of label sparsity and the lack of diversity in existing assertion data.
3) We conduct an error analysis to understand the capabilities of the best performing model on the task and to reveal directions for improvement. We make our system publicly available as a web application to allow further analyses[2].

2   Related Work

One of the earliest approaches to assertion detection is NegEx (Chapman et al., 2001), where hand-crafted word patterns are used to extract the *absent* category of assertions in discharge summaries. In 2010, the i2b2 Challenge Assertions Task (de Bruijn et al., 2011a) was introduced, and an accompanying corpus was released.

There is a variety of prior work focused on scope resolution for assertions, which differs from our setting in that it does not consider medical concepts but scopes of a certain assertion cue. Representative current approaches for this task setup include a CNN-based (Convolutional Neural Network) one by Qian et al. (2016), reaching an F1 of 0.858 on the more challenging *possible* category. Sergeeva et al. (2019) propose a LSTM-based (Long Short-Term Memory) approach to detect only *absent*

scopes. When "gold negation cues" are made available to the model and synthetic features are applied, an F1 of 0.926 is reached. NegBert (Khandelwal and Sawant, 2020) is another approach to detect *absent* scopes. As its name suggests, it is BERT-based and reaches an F1 of 0.957 on BioScope abstracts.

In contrast to these approaches we focus our work on entity-specific assertion detection, the results of which are of more practical help for supporting health professionals. Bhatia et al. (2019) explored extracting entities and negations in a joint setting, whereas the work of Harkema et al. (2009), Chen (2019) and de Bruijn et al. (2011a) is the closest to our task setup, i.e. labelling entities with an assertion class. Harkema et al. (2009) extended the NexEx algorithm with contextual properties. de Bruijn et al. (2011a) use a simple SVM classifier and Chen (2019) apply a bidirectional LSTM model with attention to the task and evaluate it on the i2b2 corpus. While these models reach F1-scores above 0.9 on the majority classes, the challenging *possible* class does not surpass 0.65. We show that medical language models outperform these scores especially regarding the minority class.

Furthermore, Wu et al. (2014) compared then state-of-the-art approaches for negation detection and found a lack of generalisation to arbitrary clinical text. We thus want to examine the transfer capabilities of recent language models to understand whether they can mitigate the phenomenon.

3   Methodology

We want to understand the abilities of medical language models on the task of assertion detection. We hence fine-tune various (medical) language models on the i2b2 corpus described below. We further apply the best performing model to the BioScope dataset and our newly introduced MIMIC-III assertion dataset without further fine-tuning to test their performance on unseen medical data.

---

[1]Annotated data available at:
`https://github.com/bvanaken/clinical-assertion-data`

[2]Demo application:
`https://ehr-assertion-detection.demo.datexis.com`

Model	F1 for		
	present	*possible*	*absent*
Earlier approaches			
SVM Classifier (de Bruijn et al., 2011b)	0.959	0.643	0.939
Conditional Softmax Shared Decoder (Bhatia et al., 2019)	–	–	0.905
Bi-directional LSTM with Attention (Chen, 2019)	0.950	0.637	0.927
Language models under evaluation			
BERT Base (Devlin et al., 2019)	0.968	0.704	0.943
BioBERT Base (Lee et al., 2020)	0.976	0.759	0.963
Bio+Clinical BERT (Alsentzer et al., 2019)	0.977	0.775	0.966
Bio+Discharge Summary BERT (Alsentzer et al., 2019)	**0.979**	**0.786**	**0.972**
Bio+Clinical Outcome Representations (CORe) (van Aken et al., 2021)	0.975	0.761	0.965
Biomed RoBERTa Base (Gururangan et al., 2020)	0.976	0.723	0.967

Table 2: Results of baseline approaches and (medical) language models on the i2b2 Assertions Task. Pre-trained medical language models outperform all earlier approaches – with a large margin on the *possible* class. Note that Bhatia et al. (2019) only evaluated their model on negation detection.

3.1 Datasets

The **2010 i2b2 Assertion Task** (de Bruijn et al., 2011a) provides a corpus of assertions in clinical discharge summaries. The task is split into six classes, namely *present, possible, absent, hypothetical, conditional* and *associated with someone else*. However, the distribution is highly skewed, such that only 6% of the assertions belong to the latter three classes. Hence we only use the *present, possible*, and *absent* assertions for our evaluation as they present the most important information for doctors.

BioScope (Vincze et al., 2008) is a corpus of assertions in biomedical publications. It was specifically curated for the study of negation and speculation (or *absent* and *possible* in this paper) scope and does not contain *present* annotations. As mentioned before, the BioScope dataset does not completely match the information need of health professionals and the i2b2 corpus lacks varied medical text types. We thus introduce a new set of labelled assertions to complement existing data.

The **MIMIC-III Clinical Database** (Johnson et al., 2016) provides texts from discharge summaries as well as other clinical notes (physician letters, nurse letters, and radiology reports) representing a promising source of varied medical text. Therefore, two annotators followed the annotation guidelines from the i2b2 challenge, and labelled 5,000 assertions, i.e. word spans of entities and their corresponding *present / possible / absent* class. The inner-annotator agreement as Cohen's kappa coefficient is **0.847**, which indicates a strong level of agreement. The annotations were further veri-

fied by a medical doctor, who provided feedback to correct a small number of labels, and confirmed that the end results were satisfactory.

It is important to note that even though the newly annotated data from MIMIC-III adds variation to the existing corpora, the dataset has its own limitations. The clinical notes are collected from a single institution (with a mostly White patient population) and from Intensive Care Unit patients only. We therefore argue that progress in assertion detection requires further initiatives for releasing more diverse sets of clinical notes.

Table 1 summarizes the assertion distribution in the introduced datasets and shows the unbalanced nature of the data.

3.2 Data Preprocessing

We make predictions about assertions on a per-entity level. However, we want our models to consider the context of an entity. We therefore pass the whole sentence to the models and surround the entity tokens with special *indicator* tokens `[entity]` whose embeddings are randomly initialised. A sample input sequence thus looks as follows: `[CLS] test results were negative for [entity] COVID-19 [entity]`.
We apply the same pre-processing to all three datasets.

3.3 Fine-tuning Medical Language Models

There are various pre-trained (bio-)medical and clinical language models available to evaluate on the assertion detection task. We select the most prevalent ones and describe them in short below:

	present	*possible*	*absent*
BioScope			
scientific pub.	–	0.593	0.845
MIMIC-III			
discharge sum.	0.951	0.663	0.939
phys. letters	0.929	0.593	0.892
nurse letters	**0.967**	**0.710**	0.900
radio. reports	0.950	0.691	**0.977**

Table 3: Experimental results (in F1) for the best performing Bio+Discharge Summary BERT model on two further assertion datasets and their different text types. Both datasets were not seen during training. Note that the number of evaluation samples is very low for some text types (i.e. *possible* class in nurse letters), which impairs the expressiveness of these results.

BERT (Devlin et al., 2019) was pre-trained on non-medical data and serves as a baseline for Transformer-base pre-trained language models. **BioBERT** (Lee et al., 2020) is a standard model for medical NLP tasks and is pre-trained on bio-medical publications. **Bio+Clinical BERT** and **Bio+Discharge Summary BERT** (Alsentzer et al., 2019) are built upon BioBERT with additional pre-training on clinical notes / discharge summaries. The **CORe** model (van Aken et al., 2021) uses BioBERT and adds a specialized clinical outcome pre-training. **Biomed RoBERTA** (Gururangan et al., 2020) is based on the RoBERTA model (Liu et al., 2019) and pre-trained on bio-medical publications. After an initial grid search we fix our hyperparameters to a learning rate of 1e-5, batch size of 32, and 2 epochs of training.

4 Evaluation and Discussion

We start by evaluating the mentioned models on the i2b2 corpus. We use training and test data as defined by in the i2b2 challenge and compare our results to previous state-of-the-art approaches in Table 2. Next, we apply the best performing Bio+Discharge Summary BERT to the BioScope and MIMIC-III corpora without additional fine-tuning (Table 3). This way we can see the model's performance on medical text from unseen sources.

4.1 Results

Language models outperform baselines. Table 2 shows that all evaluated medical language models are able to increase F1-scores on all three classes. On the most challenging *possible* class the improvement is the clearest with up to ∼15pp, which shows that the models are better in handling sparse occurrences coupled with vague expressions.

Medical pre-training is important. The vanilla BERT baseline is the weakest of our evaluated models, which shows that models specialized on the medical domain are not only effective for more complex medical tasks but also for assertion detection, which is in line with the claim by Gururangan et al. (2020) that domain-specific pre-training is almost always of use. Bio+Discharge Summary BERT is the best model – probably because it was trained on text very similar to the i2b2 corpus.

Text style matters. Table 3 shows the ability of the Bio+Discharge Summary BERT language model to transfer to other text styles. The assertions in the BioScope corpus are difficult to identify by the model as they clearly differ from the ones used by doctors in clinical notes. The text style in MIMIC-III data is more similar to the originally learned data which is reflected in the results.[3] However, physician letters appear to contain more specialized expressions and therefore evoke more errors. This points towards a lack of generalization possibly caused by the limited variety of assertion cues in the training data.

4.2 Error Analysis

We analyse all errors made by the best model to identify main sources of errors and to point towards future research directions.

Inconsistent data in pre-existing datasets account for roughly 45% of errors. This includes obvious labelling mistakes, but also disagreements among annotators. For example, phrases such as "appeared to be," "concerning for" and "consistent with" are labeled differently, as *present* or as *possible*.

Long range dependencies account for roughly 20% of all errors, in which entities and their cues have dependencies longer than a few tokens apart. While the model's attention mechanism could easily detect distant tokens, the model might have learned to only consider close assertion cues. The following is an example of a distant cue indicating the *absent* class which was missed by the model:

His <u>rash</u> on the right hand was examined further and is now <u>resolved</u>.

---

[3]Note that the model's pre-training is based on MIMIC-III and it was thus to an extent exposed to the test data. Due to the difference of the target task and the amount of total pre-training data, this influence should be negligible.

Lists of assertions are found in 8% of error samples. Here the assertion is not directly coupled to an entity but must be inferred by the way it is listed. Such somewhat ambiguous cases are usually easily understood by humans, but difficult for our models.

> <u>No</u> hydrocephalus, <u>subarachnoid hemorrhage</u>, <u>no</u> fracture.

Misspellings account for 5% of all observed errors, but they reveal a critical yet surprising limitation. For instance, the cues "appeas" and "probalbe" that indicate *possible* instances, are missed. While Transformer-based models are generally capable of dealing with misspellings due to subword tokenization, the missing variety of expressions in the data appears to let the models focus on a specific set of textual cues without generalizing to new phrases or even misspellings.

5 Conclusion and Future Work

In this work, we present an evaluation on medical language models to detect assertions in clinical texts and experimental results which show that they outperform baseline approaches. We further provided a new corpus of assertion annotations on the MIMIC-III dataset that will augment existing data collections and shows the model's capability to be transferred to other sources – if the text styles do not strongly differ. We suggest future work to investigate generalization to unseen data and expressions. We further encourage work on multi-task learning of entity extraction and assertions to support health professionals with systems that learn jointly in an end-to-end fashion.

Acknowledgments

Our work is funded by the German Federal Ministry for Economic Affairs and Energy (BMWi) under grant agreement 01MD19003B (PLASS) and 01MK2008MD (Servicemeister).

References

Emily Alsentzer, John Murphy, William Boag, Wei-Hung Weng, Di Jindi, Tristan Naumann, and Matthew McDermott. 2019. Publicly Available Clinical BERT Embeddings. In *Proceedings of the 2nd Clinical Natural Language Processing Workshop*, pages 72–78.

Parminder Bhatia, Busra Celikkaya, and Mohammed Khalilia. 2019. Joint Entity Extraction and Assertion Detection for Clinical Text. In *Proceedings of the 57th Annual Meeting of the Association for Computational Linguistics*, pages 954–959.

Wendy W Chapman, Will Bridewell, Paul Hanbury, Gregory F Cooper, and Bruce G Buchanan. 2001. A Simple Algorithm for Identifying Negated Findings and Diseases in Discharge Summaries. *Journal of Biomedical Informatics*, 34(5):301–310.

Long Chen. 2019. Attention-based Deep Learning System for Negation and Assertion Detection in Clinical Notes. *International Journal of Artificial Intelligence and Applications*, 10(1).

Berry de Bruijn, Colin Cherry, Svetlana Kiritchenko, Joel Martin, and Xiaodan Zhu. 2011a. Machine-learned Solutions for Three Stages of Clinical Information Extraction: The State of the Art at i2b2 2010. *Journal of the American Medical Informatics Association*, 18(5):557–562.

Berry de Bruijn, Colin Cherry, Svetlana Kiritchenko, Joel D. Martin, and Xiaodan Zhu. 2011b. Machine-learned solutions for three stages of clinical information extraction: the state of the art at i2b2 2010. *J. Am. Medical Informatics Assoc.*, 18(5):557–562.

Jacob Devlin, Ming-Wei Chang, Kenton Lee, and Kristina Toutanova. 2019. BERT: pre-training of deep bidirectional transformers for language understanding. In *Proceedings of the 2019 Conference of the North American Chapter of the Association for Computational Linguistics: Human Language Technologies, NAACL-HLT 2019, Volume 1 (Long and Short Papers)*, pages 4171–4186. Association for Computational Linguistics.

Suchin Gururangan, Ana Marasović, Swabha Swayamdipta, Kyle Lo, Iz Beltagy, Doug Downey, and Noah A. Smith. 2020. Don't stop pretraining: Adapt language models to domains and tasks. In *Proceedings of ACL*.

Henk Harkema, John N. Dowling, Tyler Thornblade, and Wendy Webber Chapman. 2009. Context: An algorithm for determining negation, experiencer, and temporal status from clinical reports. *J. Biomed. Informatics*, 42(5):839–851.

Alistair EW Johnson, Tom J Pollard, Lu Shen, H Lehman Li-Wei, Mengling Feng, Mohammad Ghassemi, Benjamin Moody, Peter Szolovits, Leo Anthony Celi, and Roger G Mark. 2016. MIMIC-III, a Freely Accessible Critical Care Database. *Scientific Data*, 3(1):1–9.

Aditya Khandelwal and Suraj Sawant. 2020. NegBERT: A Transfer Learning Approach for Negation Detection and Scope Resolution. In *Proceedings of The 12th Language Resources and Evaluation Conference*, pages 5739–5748.

Jinhyuk Lee, Wonjin Yoon, Sungdong Kim, Donghyeon Kim, Sunkyu Kim, Chan Ho So, and Jaewoo Kang. 2020. Biobert: a pre-trained biomedical language representation model for biomedical text mining. *Bioinform.*, 36(4):1234–1240.

Patrick S. H. Lewis, Myle Ott, Jingfei Du, and Veselin Stoyanov. 2020. Pretrained language models for biomedical and clinical tasks: Understanding and extending the state-of-the-art. In *Proceedings of the 3rd Clinical Natural Language Processing Workshop, ClinicalNLP@EMNLP 2020, Online, November 19, 2020*, pages 146–157. Association for Computational Linguistics.

Yinhan Liu, Myle Ott, Naman Goyal, Jingfei Du, Mandar Joshi, Danqi Chen, Omer Levy, Mike Lewis, Luke Zettlemoyer, and Veselin Stoyanov. 2019. Roberta: A robustly optimized BERT pretraining approach. *CoRR*, abs/1907.11692.

Yifan Peng, Shankai Yan, and Zhiyong Lu. 2019. Transfer learning in biomedical natural language processing: An evaluation of BERT and elmo on ten benchmarking datasets. In *Proceedings of the 18th BioNLP Workshop and Shared Task, BioNLP@ACL 2019, Florence, Italy, August 1, 2019*, pages 58–65. Association for Computational Linguistics.

Sujan Perera, Amit Sheth, Krishnaprasad Thirunarayan, Suhas Nair, and Neil Shah. 2013. Challenges in Understanding Clinical Notes: Why NLP Engines Fall Short and Where Background Knowledge Can Help. In *Proceedings of the 2013 International Workshop on Data Management & Analytics for Healthcare*, page 21–26.

Zhong Qian, Peifeng Li, Qiaoming Zhu, Guodong Zhou, Zhunchen Luo, and Wei Luo. 2016. Speculation and Negation Scope Detection via Convolutional Neural Networks. In *Proceedings of the 2016 Conference on Empirical Methods in Natural Language Processing*, pages 815–825.

Elena Sergeeva, Henghui Zhu, Peter Prinsen, and Amir Tahmasebi. 2019. Negation Scope Detection in Clinical Notes and Scientific Abstracts: A Feature-enriched LSTM-based Approach. *AMIA Summits on Translational Science Proceedings*, 2019:212.

Betty van Aken, Jens-Michalis Papaioannou, Manuel Mayrdorfer, Klemens Budde, Felix A. Gers, and Alexander Löser. 2021. Clinical outcome prediction from admission notes using self-supervised knowledge integration. In *Proceedings of the 16th Conference of the European Chapter of the Association for Computational Linguistics, EACL 2021*. Association for Computational Linguistics.

Veronika Vincze, György Szarvas, Richárd Farkas, György Móra, and János Csirik. 2008. The BioScope Corpus: Biomedical Texts Annotated for Uncertainty, Negation and Their Scopes. *BMC bioinformatics*, 9(11):1–9.

Stephen Wu, Timothy Miller, James Masanz, Matt Coarr, Scott Halgrim, David Carrell, and Cheryl Clark. 2014. Negation's not solved: generalizability versus optimizability in clinical natural language processing. *PLoS One*, 11(9).

Extracting Appointment Spans from Medical Conversations

Nimshi Venkat Meripo
Abridge AI Inc.
venkatm@abridge.com

Sandeep Konam
Abridge AI Inc.
san@abridge.com

Abstract

Extracting structured information from medical conversations can reduce the documentation burden for doctors and help patients follow through with their care plan. In this paper, we introduce a novel task of extracting appointment spans from medical conversations. We frame this task as a sequence tagging problem and focus on extracting spans for appointment reason and time. However, annotating medical conversations is expensive, time-consuming, and requires considerable domain expertise. Hence, we propose to leverage weak supervision approaches, namely incomplete supervision, inaccurate supervision, and a hybrid supervision approach and evaluate both generic and domain-specific, ELMo, and BERT embeddings using sequence tagging models. The best performing model is the domain-specific BERT variant using weak hybrid supervision and obtains an F1 score of 79.32.

1 Introduction

Increased Electronic Health Records (EHR) documentation burden is one of the leading causes for physician burnout (Downing et al., 2018; Collier, 2017). Although EHRs facilitate effective workflow and access to data, several studies have shown that physicians spend more than half of their workday on EHRs (Arndt et al., 2017). This leads to decreased face time with patients and reduced work satisfaction for physicians (Drossman and Ruddy, 2019; Sinsky et al., 2016). For these reasons, there has been growing interest in using machine learning tecniques to extract relevant information for a medical record from medical conversations (Lin et al., 2018; Schloss and Konam, 2020).

On the other hand, research shows that approximately 23% of patients do not show up for their doctor appointments (Dantas et al., 2018). Missed appointments have a large impact on hospitals' ability to provide efficient and effective services (Chandio et al., 2018). Studies in Callen et al. (2012)

DR: And the thing with angiogram, uh, um, um, we already scheduled, uh, the days I do it is Mondays and on Fridays. That's the best time for me to do the angiogram.
DR: Um, yeah, I, I guess I could make it Monday next week.
DR: Or next, or Friday?
DR: Which one is better for you?
PT: Um, well, it's already - Today is, uh, uh, Tuesday.
DR: Yeah, let's do it as soon as possible.
PT: Okay, let's do it on Monday.

Figure 1: An utterance window from a medical conversation annotated with appointment reason and time spans.

also show that a significant number of patients miss their lab appointments. Missed lab appointments can put a patient's health at risk and allow diseases to progress unnoticed (Mookadam et al., 2016). One of the main reasons for no-shows is patient forgetfulness (Ullah et al., 2018). Mookadam et al. (2016) and Perron et al. (2013) show that proactive reminders through text messages, calls, and mobile applications are promising and significantly decrease the missed appointment rates.

In line with the aforementioned value, appointment span extraction from medical conversations can help physicians document the care plan regarding diagnostics (Dx), procedures (Px), follow-ups, and referrals. It can also directly impact a patient's ability to keep their appointments. In this work, we investigate extracting the appointment reason and time spans from medical conversations as shown in Figure 1. The reason span refers to a phrase that corresponds to Dx, Px, follow-ups and referrals.

Proceedings of the Second Workshop on Natural Language Processing for Medical Conversations, pages 41–46
July 6, 2021. ©2021 Association for Computational Linguistics

The time span refers to a phrase that corresponds to the time of the appointment. To tackle this task, we collected a dataset for both reason and time spans and framed it as a sequence tagging problem. Our contributions include: (i) defining the appointment span extraction task, (ii) describing the annotation methodology for labeling the medical conversations, (iii) investigating weak supervision approaches on sequence tagging models using both generic and domain-specific ELMo and BERT embeddings, and (iv) performing error analysis to gather insights for improving the performance.

2 Related work

Extracting Dx, Px and time expressions has been the subject of past work. Tools such as Clinical Text Analysis and Knowledge Extraction System (cTAKES) (Savova et al., 2010) and MetaMa (Aronson, 2006) are widely used in the biomedical field to extract medical entities. Both of these tools use the Unified Medical Language System (UMLS) (Bodenreider, 2004) to extract and standardize medical concepts. However, UMLS is primarily designed for written clinical text, not for spoken medical conversations. Further, research on date-time entity extraction from text is task agnostic. Rule-based approaches like HeidelTime (Strötgen and Gertz, 2010), and SUTime (Chang and Manning, 2012) mainly handcraft rules to identify time expression in the text. Learning-based approaches typically extract features from text and apply statistical models such as Conditional Random Fields (CRFs). While these tools perform well for generic clinical and date-time entity extraction from texts, they don't fare as well on task-specific entity extraction, where only a subset of the entities present in the text is relevant to solving the task.

Recently, there has been an increasing interest in medical conversations-centered applications (Chiu et al., 2017). Du et al. (2019a,b) proposed several methods for extracting entities such as symptoms and medications and their relations. Selvaraj and Konam (2019) and Patel et al. (2020) examined the task of medication regimen extraction. While recent research in medical conversations is primarily focused on extracting symptoms and medications, we propose a new task of extracting appointment spans. Our framing of this task as a sequence tagging problem is similar to Du et al. (2019a,b); however, they use a fully supervised approach and mainly focus on relation extraction, whereas we

investigate weak supervision for appointment span extraction. Moreover, we evaluate both generic and domain-specific ELMo and BERT models in our task.

3 The Appointment Span Extraction Task

3.1 Corpus Description

Our corpus consists of human-written transcripts of 23k fully-consented and manually de-identified real doctor-patient conversations. Each transcript is annotated with utterance windows where the appointment is discussed. We have obtained a total of 43k utterance windows that discuss appointments. Of the 43k utterance windows, 3.2k utterances windows from 5k conversations are annotated with two types of spans: appointment reason and appointment time (Figure 1). We have also obtained annotations for other span types such as appointment duration and frequency, however due to infrequency of such spans, we have not included these spans in this study.

3.2 Annotation Methodology

Span Type	Examples
Reason	follow-up, dermatologist, MRI, chemotherapy, chemo, physical, heart surgery
Time	about a month, every two weeks, in the middle of August, July 2021, before the next appointment

Table 1: Examples of annotated spans.

A team of 15 annotators annotated the dataset. The annotators were highly familiar with medical language and have significant experience in medical transcription and billing. We have distributed 3.2k utterance windows equally among 15 annotators. Each utterance window is doubly-annotated with appointment spans, and the authors resolved any conflicting annotations. We collect the spans of text describing the reason and time for only future appointments. We show examples of reason and time spans in Table 1. Overall, 6860 reason spans and 2012 time spans are annotated, and the average word lengths for reason and time spans are 1.6 and 2.3, respectively.

Reason span   The reason span captures four types of appointments: follow-ups, referrals, diagnostics, and procedures. Phrases of body parts and substances are also captured if they are mentioned in relation to the appointment reason (e.g., ultrasound of my *kidney*, surgery for the *heart valve*). We also annotated the spans where the appointment reason is expressed in informal language (e.g., *see you back* for follow-ups, let's do your *blood* for a blood test).

Time span   The time span captures the time of an appointment. We also included prepositions (eg. *in* two days, *at* 3 o'clock) and time modifiers (eg., *after* a week, *every* year) in this span. In cases where multiple different time phrases are present for an appointment, annotators were instructed to annotate a time phrase that is confirmed by either patient or doctor, or annotate potentially valid time phrases if the discussion is ambiguous.

Due to the conversational nature, appointment reason and time are often discussed multiple times using the same phrase or a synonymous phrase (e.g., a *blood test* called *FibroTest*, *Monday* or *Monday next week*). To maintain consistency across different conversations, annotators were instructed to mark all occurrences of the span.

3.3   Methods

To account for the limited set of annotations, we employed weak supervision approaches. We specifically used inaccurate supervision, incomplete supervision (Zhou, 2018) and developed a hybrid approach that utilizes both inaccurate and incomplete supervision.

Inaccurate Supervision   Inaccurate supervision is a scenario where the training labels provided are not always the ground-truth; in other words, the training labels suffer from errors. We take advantage of off-the-shelf tools such as UMLS and spaCy (Honnibal et al., 2020) to automatically annotate reason and time spans. For the reason span, we perform a dictionary lookup in UMLS vocabularies and extract any span with a semantic type belonging to Dx, Px, and body parts. For the time span, we use spaCy's named entity recognition (NER) model to extract spans belonging to time and date. To reduce the inaccuracies, we included only the utterance windows with at least one reason phrase and one time phrase. Using this approach, we ob-

tained 20k utterance windows with both appointment reason and time spans.

Incomplete Supervision   Incomplete Supervision refers to a scenario where only a small subset of data has annotated labels. For this scenario, we use 2.5k conversations from manual span annotated corpus conversations, which resulted in 1292 utterance windows.

Hybrid Supervision   In this approach, we apply both inaccurate and incomplete supervision techniques sequentially. To avoid catastrophic forgetting (McCloskey and Cohen, 1989), the models are first trained with inaccurate supervision and then fine-tuned with incomplete supervision.

We use a 85:15 split of the remaining 1844 manual span annotated utterance windows for testing and validation purposes. To make the test dataset more difficult, we used a weighted sampling technique in which each appointment span is weighted by the inverse probability of it being sampled.

4   Models

In this section, we briefly describe our two models that use variants of contextualized embeddings namely, ELMo (Peters et al., 2018) and BERT (Devlin et al., 2018).

4.1   ELMo-based CRF

Our model is a 2-layer BiLSTM network using GloVe word embeddings and a character-based CNN representation trained with CRF loss. Similar to the approach taken in Peters et al. (2018), our model is enhanced by concatenating a weighted average of ELMo embeddings with GloVe and character embeddings. We next describe the two variants of ELMo models we use.

ELMo   The original ELMo model is pre-trained on generic language corpora using the 1 Billion Words dataset (Chelba et al., 2013).

BioELMo   BioELMo (Jin et al., 2019) is a biomedical variant of ELMo trained on 10M recent abstracts (2.46B tokens) from PubMed.

4.2   BERT-based classifier

Similar to the approach taken in Devlin et al. (2018), we use a token level classifier instead of a CRF layer and fine-tune variants of the BERT model. We next describe the variants of BERT models we use.

BERT   The original BERT model is trained on BooksCorpus (Zhu et al., 2015) and English Wiki.

BioBERT   BioBERT (Lee et al., 2019) further pre-trains the BERT-base model on a large corpus of PubMed abstracts containing 4.5B words.

4.3   Experiment details

Model	Embedding Size	Learning rate
ELMo variants	1024	1e-3
BERT variants	768	3e-5

Table 2: Experiment configurations for the models.

The experiment configuration for ELMo and BERT variants used in our experiments is shown in Table 2. Both ELMo and BERT variants use an uncased vocabulary. The span labels are represented using the IOB2 tagging scheme (Sang and Veenstra, 1999).

5   Evaluation

To evaluate our models, we measure micro-averaged Precision, Recall, and F1 of reason and time spans on the test dataset (Table 3). Both ELMo and BERT variants performed similarly with inaccurate supervision owing to the noisy nature of the inaccurate supervision. With the incomplete supervision approach, the performance improved considerably, ranging from 49% in ELMo to 60% in BioBERT. Both BioELMo and BioBERT gained more than the ELMo and BERT variants, respectively. However, with hybrid supervision, both the ELMo variants benefited most and achieved similar performance nullifying the advantage of the in-domain pre-training of BioELMo.

On the other hand, the BERT variants showed a minor improvement with hybrid supervision. The BERT variants consistently performed better than ELMo variants, and the domain-specific pre-training has only a minor impact on BERT when compared to ELMo. Overall, the proposed hybrid supervision approach has consistently improved performance across all model variants and the results show that augmenting the training data with inaccurate supervision can improve the performance.

In order to assess performance at each span type, we chose the best performing BioBERT-hybrid model. For both span types precision was lower than recall (Table 4) suggesting a higher percentage of false positives than false negatives.

Model	P	R	F1
ELMo-inaccurate	58.76	43.29	49.85
ELMo-incomplete	71.76	78.03	74.77
ELMo-hybrid	77.05	77.22	77.14
BioELMo-inaccurate	58.09	42.44	49.05
BioELMo-incomplete	73.30	78.19	75.67
BioELMo-hybrid	74.74	79.69	77.14
BERT-inaccurate	58.95	42.73	49.54
BERT-incomplete	73.96	**82.29**	77.91
BERT-hybrid	76.16	81.44	78.71
BioBERT-inaccurate	58.62	42.41	49.22
BioBERT-incomplete	76.98	80.66	78.77
BioBERT-hybrid	**77.23**	81.53	**79.32**

Table 3: Evaluation of weak supervision methods; P: Precision, R: Recall, F1: F1 score.

Span Type	Precision	Recall	F1	# Occurences
Reason	80.52	84.27	82.36	3459 (3687)
Time	66.24	72.02	69.01	997 (1163)

Table 4: Performance of BioBERT-hybrid model and the number of occurrences of each span type in ground truths and predictions respectively.

6   Error Analysis

Error Type	Reason	Time
Correct Label - Overlapping Span	6.83	14.61
Wrong Label - Correct Span	0.08	0.08
Wrong Label - Overlapping Span	0.13	0.77
Complete False Positive	13.77	23.12
Complete False Negative	8.03	11.41

Table 5: Percentage of error types on the test set using the BioBERT-hybrid model.

To better understand the errors in predictions, we computed percentages of different types of errors (Table 5). The cases where the model predicted the right label but with an overlapping span (Correct Label-Overlapping Span) are mainly due to inconsistencies in annotations. The primary source of these inconsistencies is when annotators missed annotating a prepositional phrase or a time modifier phrase in the time span. Wrong label errors (Wrong Label - Correct Span, Overlapping Span) are minimal, suggesting that the model distinguishes between the time and reason spans very well.

Complete false positives and false negatives are the significant sources of errors for both reason and time spans and our qualitative analysis suggests that these cases often happen when multiple reason phrases and time phrases are present in the utterance window, but only a subset of them are valid. Because the task actually involves two different aspects, extracting reason and time mentions and spotting their confirmation clues, it may be difficult for the trained system to select exactly the confirmed reason or time mentions without explicitly modeling their relations. The ambiguity due to the oral nature of the conversations also makes it difficult to spot the confirmation clues.

Notably, we observe that the portion of complete false positives for the time span is significantly higher than reason spans. For example, the conversation in Figure 1 discusses several options for the appointment time, but the patient finally settles for Monday. The model often struggles with such cases and also extracts time mentions that are not confirmed. Using SpaCy's NER, we find that 87% of these errors occurred when multiple time phrases are present, but not all are valid. The model may have difficulty with these cases because they amount to only 21.3% of the manually annotated time spans. Further, the annotated time spans are infrequent by a factor of three than the reason spans. These reasons explain why the F1 score on time span is significantly lower than the reason span.

7 Conclusion

In summary, we defined a novel task of extracting appointment spans from medical conversations, described our annotation methodology, and employed three weak supervision approaches to account for the limited set of annotations. Our proposed hybrid weak supervision approach showed improvement across all our experiments. Finally, our error analysis shows that a significant portion of the errors comes from false positives where the model has difficulty in identifying the correct span when multiple appointment reason or time mentions are present. In future work, we plan to study the data augmentation approaches as well as joint entity and relation extraction approaches to improve performance on difficult examples. We also plan to study the generalization of this work to automatic transcripts, whose transcription error rate may challenge entity detection.

References

B. G. Arndt, J. W. Beasley, M. D. Watkinson, J. L. Temte, W. J. Tuan, C. A. Sinsky, and V. J. Gilchrist. 2017. Tethered to the EHR: Primary Care Physician Workload Assessment Using EHR Event Log Data and Time-Motion Observations. *Ann Fam Med*, 15(5):419–426.

Alan R Aronson. 2006. Metamap: Mapping text to the umls metathesaurus. *Bethesda, MD: NLM, NIH, DHHS*, 1:26.

Olivier Bodenreider. 2004. The unified medical language system (umls): integrating biomedical terminology. *Nucleic acids research*, 32(suppl_1):D267–D270.

Joanne L Callen, Johanna I Westbrook, Andrew Georgiou, and Julie Li. 2012. Failure to follow-up test results for ambulatory patients: a systematic review. *Journal of general internal medicine*, 27(10):1334–1348.

A Chandio, Z Shaikh, K Chandio, SM Naqvi, and SA Naqvi. 2018. Can "no shows" to hospital appointment be avoided. *Clin Surg*, 3:1918.

Angel X Chang and Christopher D Manning. 2012. Sutime: A library for recognizing and normalizing time expressions. In *Lrec*, volume 3735, page 3740.

Ciprian Chelba, Tomas Mikolov, Mike Schuster, Qi Ge, Thorsten Brants, Phillipp Koehn, and Tony Robinson. 2013. One billion word benchmark for measuring progress in statistical language modeling. Technical report, Google.

Chung-Cheng Chiu, Anshuman Tripathi, Katherine Chou, Chris Co, Navdeep Jaitly, Diana Jaunzeikare, Anjuli Kannan, Patrick Nguyen, Hasim Sak, Ananth Sankar, Justin Tansuwan, Nathan Wan, Yonghui Wu, and Xuedong Zhang. 2017. Speech recognition for medical conversations. *CoRR*, abs/1711.07274.

Roger Collier. 2017. Electronic health records contributing to physician burnout.

Leila F Dantas, Julia L Fleck, Fernando L Cyrino Oliveira, and Silvio Hamacher. 2018. No-shows in appointment scheduling—a systematic literature review. *Health Policy*, 122(4):412–421.

Jacob Devlin, Ming-Wei Chang, Kenton Lee, and Kristina Toutanova. 2018. Bert: Pre-training of deep bidirectional transformers for language understanding. *arXiv preprint arXiv:1810.04805*.

N Lance Downing, David W Bates, and Christopher A Longhurst. 2018. Physician burnout in the electronic health record era: are we ignoring the real cause?

D. A. Drossman and J. Ruddy. 2019. Improving Patient-Provider Relationships to Improve Health Care. *Clin. Gastroenterol. Hepatol.*

Nan Du, Kai Chen, Anjuli Kannan, Linh Tran, Yuhui Chen, and Izhak Shafran. 2019a. Extracting symptoms and their status from clinical conversations. *arXiv preprint arXiv:1906.02239*.

Nan Du, Mingqiu Wang, Linh Tran, Gang Li, and Izhak Shafran. 2019b. Learning to infer entities, properties and their relations from clinical conversations. *arXiv preprint arXiv:1908.11536*.

Matthew Honnibal, Ines Montani, Sofie Van Landeghem, and Adriane Boyd. 2020. spaCy: Industrial-strength Natural Language Processing in Python.

Qiao Jin, Bhuwan Dhingra, William W Cohen, and Xinghua Lu. 2019. Probing biomedical embeddings from language models. *arXiv preprint arXiv:1904.02181*.

Jinhyuk Lee, Wonjin Yoon, Sungdong Kim, Donghyeon Kim, Sunkyu Kim, Chan Ho So, and Jaewoo Kang. 2019. BioBERT: a pretrained biomedical language representation model for biomedical text mining. *Bioinformatics*, 36(4):1234–1240.

Steven Y Lin, Tait D Shanafelt, and Steven M Asch. 2018. Reimagining clinical documentation with artificial intelligence. In *Mayo Clinic Proceedings*, volume 93, pages 563–565. Elsevier.

Michael McCloskey and Neal J Cohen. 1989. Catastrophic interference in connectionist networks: The sequential learning problem. In *Psychology of learning and motivation*, volume 24, pages 109–165. Elsevier.

Martina Mookadam, Michael Grover, Chris Pullins, Mary Winscott, and Susan Pierce. 2016. Simple interventions improve the quality of a missed lab appointment process. *BMJ Open Quality*, 5(1).

Dhruvesh Patel, Sandeep Konam, and Sai P. Selvaraj. 2020. Weakly supervised medication regimen extraction from medical conversations.

Noelle Junod Perron, Melissa Dominicé Dao, Nadia Camparini Righini, Jean-Paul Humair, Barbara Broers, Françoise Narring, Dagmar M Haller, and Jean-Michel Gaspoz. 2013. Text-messaging versus telephone reminders to reduce missed appointments in an academic primary care clinic: a randomized controlled trial. *BMC health services research*, 13(1):1–7.

Matthew Peters, Mark Neumann, Mohit Iyyer, Matt Gardner, Christopher Clark, Kenton Lee, and Luke Zettlemoyer. 2018. Deep contextualized word representations. In *Proceedings of the 2018 Conference of the North American Chapter of the Association for Computational Linguistics: Human Language Technologies, Volume 1 (Long Papers)*, pages 2227–2237, New Orleans, Louisiana. Association for Computational Linguistics.

Erik F Sang and Jorn Veenstra. 1999. Representing text chunks. *arXiv preprint cs/9907006*.

Guergana K Savova, James J Masanz, Philip V Ogren, Jiaping Zheng, Sunghwan Sohn, Karin C Kipper-Schuler, and Christopher G Chute. 2010. Mayo clinical text analysis and knowledge extraction system (ctakes): architecture, component evaluation and applications. *Journal of the American Medical Informatics Association*, 17(5):507–513.

Benjamin Schloss and Sandeep Konam. 2020. Towards an automated soap note: Classifying utterances from medical conversations. In *Machine Learning for Healthcare Conference*, pages 610–631. PMLR.

Sai P. Selvaraj and Sandeep Konam. 2019. Medication regimen extraction from medical conversations.

Christine Sinsky, Lacey Colligan, Ling Li, Mirela Prgomet, Sam Reynolds, Lindsey Goeders, Johanna Westbrook, Michael Tutty, and George Blike. 2016. Allocation of Physician Time in Ambulatory Practice: A Time and Motion Study in 4 Specialties. *Annals of Internal Medicine*, 165(11):753–760.

Jannik Strötgen and Michael Gertz. 2010. Heideltime: High quality rule-based extraction and normalization of temporal expressions. In *Proceedings of the 5th International Workshop on Semantic Evaluation*, pages 321–324.

S Ullah, S Rajan, T Liu, E Demagistris, R Jahrstorfer, S Anandan, C Gentile, and A Gil. 2018. Why do patients miss their appointments at primary care clinics. *J Fam Med Dis Pre*, 4:09.

Zhi-Hua Zhou. 2018. A brief introduction to weakly supervised learning. *National science review*, 5(1):44–53.

Yukun Zhu, Ryan Kiros, Rich Zemel, Ruslan Salakhutdinov, Raquel Urtasun, Antonio Torralba, and Sanja Fidler. 2015. Aligning books and movies: Towards story-like visual explanations by watching movies and reading books. In *Proceedings of the IEEE international conference on computer vision*, pages 19–27.

Building Blocks of a Task-Oriented Dialogue System in the Healthcare Domain

Heereen Shim[1,2,3], Dietwig Lowet[1], Bart Vanrumste[2,3] and Stijn Luca[4]
[1]Philips Research, Eindhoven, the Netherlands
[2]Campus Group T, e-Media Research Lab, KU Leuven, Leuven, Belgium
[3]Department of Electrical Engineering (ESAT), STADIUS, KU Leuven, Leuven, Belgium
[4]Department of Data Analysis and Mathematical Modelling, Ghent University, Ghent, Belgium
`{heereen.shim, dietwig.lowet}@philips.com`
`{bart.vanrumste}@kuleuven.be`
`{stijn.luca}@ugent.be`

Abstract

There has been significant progress in dialogue systems research. However, dialogue systems research in the healthcare domain is still in its infancy. In this paper, we analyse recent studies and outline three building blocks of a task-oriented dialogue system in the healthcare domain: i) privacy-preserving data collection; ii) medical knowledge-grounded dialogue management; and iii) human-centric evaluations. To this end, we propose a framework for developing a dialogue system and show preliminary results of simulated dialogue data generation by utilising expert knowledge and crowdsourcing.

1 Introduction

There has been significant progress in the research field of the dialogue system in past years with the help of large-scale pre-trained language models (LMs) (Vaswani et al., 2017; Radford et al., 2019; Lewis et al., 2020). Pre-trained LMs show a good generalised ability obtained from massive training data collected from the internet and achieve state-of-the-art performance over a wide range of dialogue domains (Zhang et al., 2020). While many studies exist on general purpose dialogues, the research on dialogue systems for healthcare applications is still in its infancy.

There are two major directions in the development of a dialogue system. One direction is to build a chatbot that can have a conversation with a user. This approach mainly focuses on generating appropriate response given user input and dialogue history. Researchers have been working on this direction to create systems to produce more human-like (Adiwardana et al., 2020), consistent (Wolf et al., 2019), and empathetic (Rashkin et al., 2019) responses. The other direction is to build a task-oriented dialogue system that performs a specific task, such as triage or diagnosis within the healthcare domain where researchers focus on developing systems that can detect implicit symptoms or make precise diagnosis/triage result (Middleton et al., 2016; Razzaki et al., 2018; Xu et al., 2019; Wei et al., 2018).

In this study, we consider a dialogue system for a sleep coaching programme for healthy people who would like to optimise their sleep. Motivated by cognitive behaviour therapy for insomnia (CBT-I), we focus on investigating the relationship between how people think, behave, and sleep (Morin et al., 2006). The first step of the coaching programme is a complaints assessment to identify sleep issues and their potential causes and decide the next step (e.g., referring to sleep apnea treatment, providing a sleep education, suggesting a behaviour change programme, etc). During this process, a coaching provider (coach) plays as an active listener, asking questions to probe specific information, while a coaching receiver (user) has more chance to provide complaints and elaborate on these.

Real challenges in the development of a dialogue system, especially a machine learning-based system, come from three fundamental questions: i) how to obtain relevant data; ii) how to develop an automated system; and iii) how to evaluate a system. In this paper, we first analyse existing approaches that address the above questions (Section 2). Then we propose our method to address these questions (Section 3) and show preliminary results and discuss its limitations (Section 4).

The major contributions of this paper are as follows:

- Identifying gaps in existing dialogue systems in the healthcare domain.

- Proposing a framework consisting of three

47

Proceedings of the Second Workshop on Natural Language Processing for Medical Conversations, pages 47–57
July 6, 2021. ©2021 Association for Computational Linguistics

building blocks.

- Constructing a dataset to illustrate the validity of the proposed method.

2 Related Work

2.1 Data Collection

Obtaining dialogue data is time-consuming and might not be available, especially in the healthcare domain. There are several recent studies on creating a large-scale conversation dataset in the healthcare domain by scrapping dialogues from online websites (Wei et al., 2018; Xu et al., 2019; Zeng et al., 2020). These web-scraping approaches, however, are not scalable and might create potential privacy issues.

To mitigate the scalability issue, some studies leverage domain knowledge to generate simulated dialogue. For example, Liednikova et al. (2020) modelled a typical dialogue flow between doctor-patient in the form of a tree. Then they augmented data by adding similar sentences extracted from an online forum. A drawback of this approach is that access to data sources is required and it might not be available within European countries in the light of the General Data Protection Regulation (GDPR). Contrary to this, Liu et al. (2019) proposed a framework for generating simulated data based on templates, which are logically and clinically verified, and incorporated linguistic knowledge to create diverse augmented data.

Another line of work on collecting dialogue data is to utilise a user simulator. User simulator has been widely used to interact with a dialogue system (Shi et al., 2019). Some of the recent works adapted agenda-based user simulator (Schatzmann and Young, 2009) to create training data for dialogue-based diagnosis systems (Wei et al., 2018; Xu et al., 2019). However, they still utilised web-scrapped data to model user behaviour.

2.2 Dialogue Management

Dialogue management is a component of a dialogue system that processes dialogue context and decides the right next action for the agent to take (Young et al., 2013). For health-related dialogue (e.g., symptom check, triage, diagnosis, etc), the role of dialogue management is to decide what to ask, answer, or inform given the context.

Middleton et al. (2016) casts triage into a sequence of questions and answers. They modelled triage flow as a graph by encoding medical knowledge. This graph plays the role of dialogue management to guide a system to interact with users and make a triage decision. This approach has the following advantages: 1) it alleviates the issue of data collection since they do not rely on machine learning with large-scale data but human expert knowledge; 2) it can reason about its predictions. However, the limitation of this approach is that it requires a lot of expert resources.

Some task-oriented dialogue systems learn how to manage a dialogue flow by reinforcement learning (RL) (Wei et al., 2018; Xu et al., 2019). For example, Wei et al. (2018) framed a dialogue management module as an RL agent with a deep Q-network (Mnih et al., 2015). With this approach, the RL agent can decide the next action (i.e., to inquire about implicit symptoms, to make a diagnosis, etc) based on the current dialogue state. Later, Xu et al. (2019) showed that incorporating a medical knowledge graph and symptom-disease relations can allow an RL agent to ask more relevant implicit symptoms and make a precise diagnosis.

There are also some recent works on developing generative models for an end-to-end dialogue system in the healthcare domain (Liednikova et al., 2020; Zeng et al., 2020) by utilising generative pre-trained LMs (Wolf et al., 2019; Radford et al., 2018, 2019; Lewis et al., 2020; Zhang et al., 2020; Vaswani et al., 2017). However, considering the fact that these generative models are less controllable (Wallace et al., 2019; Sheng et al., 2019), using a pre-trained LM-based generative model for health-related conversation could be risky.

2.3 Evaluation

To evaluate a task-oriented dialogue system, multiple metrics are used; both automatic evaluation metrics and human evaluation metrics. Automatic evaluation metrics include success rate, the average number of turns per dialogue session, matching rate, and average reward for an RL-based system (Li et al., 2017; Wei et al., 2018; Xu et al., 2019). While the automated metrics focus on task completion, human evaluation metrics consider qualitative aspects of the dialogue, such as the quality of dialogue flow, the appropriateness of decision making (diagnosis validity), and dialogue fluency scored by experts (Razzaki et al., 2018; Xu et al., 2019).

However, user perspective has been less considered in evaluating a task-oriented dialogue system in healthcare. User-centric metric, such as a user rating score or user preference score (Li et al., 2019), is widely used for evaluating general-purpose dialogue systems (Shi et al., 2019; Shah et al., 2018; Budzianowski and Vulić, 2019; Roller et al., 2020). A user-centric metric can not only be used to assess the performance of a system but debug a system as well. For example, a user might have difficulty understanding the complex language that a system uses or be annoyed by too many questions without a proper explanation. In this case, using proper user-centric metrics can provide an insight into which aspects of a system should be updated.

3 Building Blocks

Here we outline three building blocks of a dialogue system in the healthcare domain and identify open research questions for each building block. To this end, we propose a framework for developing a conversation agent for healthcare-related dialogues.

3.1 Privacy-Preserving Data Collection

As mentioned earlier, the potential privacy issues create challenges in data collection, especially in European countries in the light of GDPR. We identify three potential methods of data collection while safeguarding privacy. The first potential method is to apply appropriate privacy protection techniques to the collected data, such as de-identification that replaces the sensitive information for text (Neamatullah et al., 2008; Meystre et al., 2010; Neubauer and Heurix, 2011). The second potential method is to generate synthetic data by training generative models on the collected data (Guan et al., 2019; Hatua et al., 2019; Pan et al., 2020). The third potential method is to generate simulated data by building a user simulator that can interact with a dialogue system (Wei et al., 2018; Xu et al., 2019; Kao et al., 2018). Applying these three methods, however, entails the following consideration: How much is the risk of information leakage? What is the difference in performance between models trained on de-identified, synthesised, simulated and real data?

3.2 Medical Knowledge-Grounded Dialogue Management

Unlike an open-domain dialogue, healthcare-related dialogue should be grounded in medical knowledge. Two types of knowledge can be in-cluded in a dialogue system. The first type of knowledge is the knowledge about dialogue between healthcare professional and healthcare recipient. For example, in the healthcare domain, there exists a typical structure of dialogue that is advised to be followed. Modelling a dialogue structure can guide a system to have an appropriate dialogue flow (Middleton et al., 2016; Razzaki et al., 2018). The second type of knowledge is medical knowledge, including correlations between symptoms and causal relation between symptom and diseases. Incorporating medical knowledge can allow a system to have more appropriate dialogue and make a precise decision (Ni et al., 2017; Ghosh et al., 2018; Chen et al., 2020; Xu et al., 2019). The open questions are: How to efficiently encode expert knowledge into a machine-accessible format (e.g., knowledge graph, knowledge base) and how to incorporate it into a machine learning model? How to maintain the previously built knowledge to keep updated?

3.3 Human-Centric Evaluation

Since a dialogue system is designed to interact with a user, a human evaluation should be is considered as an ideal evaluation. More specifically, two types of human evaluations metrics should be considered to correctly evaluate a dialogue system in the healthcare domain: one from the expert (healthcare professional) perspective and the other from the end-user (healthcare recipient) perspective. Experts from the domain should validate the appropriateness of the dialogue actions made by an agent and assess the quality of the dialogue (Razzaki et al., 2018; Xu et al., 2019). Also, end-user should evaluate a system in terms of satisfaction, usability, and comprehensibility by rating each aspect (Shi et al., 2019; Shah et al., 2018) or deciding the preferred system (Li et al., 2019; Roller et al., 2020). This is associated with the following questions: Which aspects are critical to assess both the functionality and the usability of a system? How can these evaluations be reflected to update a system efficiently?

3.4 A Proposed Framework

Considering the above-mentioned building blocks, we propose a framework for developing a conversational agent in the healthcare domain as illustrated in Figure 1.

Simulated Data Generation The proposed framework generates simulated dialogue data to

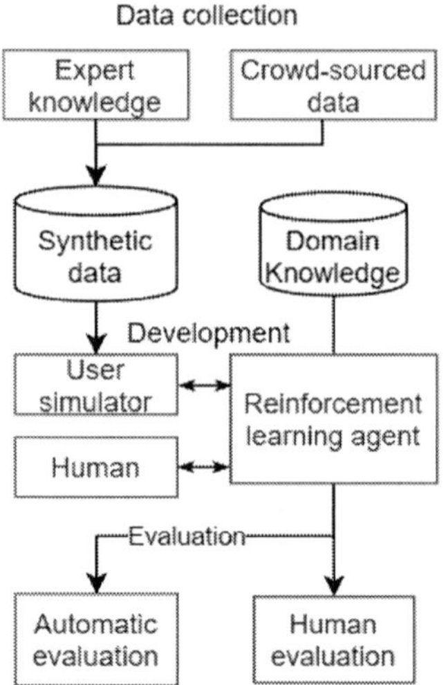

Figure 1: Overview of the proposed framework.

avoid potential privacy issue in data collection. We follow recent works on generating a simulated data set based on the knowledge of user behaviour and the characteristics of dialogue without using real user data (Shah et al., 2018). This consists of two steps: firstly, a template is constructed by exploiting expert knowledge. Secondly, data is augmented by utilising crowdsourcing.

Reinforcement Learning Agent Similar to previous studies (Wei et al., 2018; Xu et al., 2019), we frame a dialogue management module as an RL agent. We propose a two-step training procedure. At the first step, the RL agent is trained with a user simulator, either an agenda-based (Schatzmann and Young, 2009) or a model-based (El Asri et al., 2016; Kreyssig et al., 2018) one. At the second step, the RL agent is further trained by interacting with real-world users.

Model evaluation To evaluate the model, we use both an automatic evaluation metric and a human evaluation metric. Since we consider a task-oriented dialogue system, success rate and matching rate (Xu et al., 2019) are used as automatic metrics. For the human evaluation metric, validity scores by experts (Razzaki et al., 2018) and preference scores by users (Li et al., 2019) are used.

4 Preliminary Results

This section describes an initial approach of generating simulated dialogues based on a template and crowdsourced data. The goal of a dialogue is to assess user complaints related to their sleep and

identify all potential behavioural factors that might be associated with the reported complaints.

4.1 Dialogue Template

We consulted an expert in the sleep domain to model a dialogue between user and coach in the form of a tree. The dialogue template is structured in three parts of questions and potential answers related to sleep issues, the impacts of sleep issues, and behavioural factors (i.e., habits/lifestyles that might affect sleep quality). More specifically, one open-ended question that is associated with 11 potential answers and two close-ended follow-up questions (i.e., the frequency and the duration of the reported issue) in the sleep issue part, one open-ended question that is associated with 10 potential answers and one close-ended follow-up question (i.e., an enquiry regarding daytime fatigue) in the impact part, and 11 close-ended questions in the behavioural factor part. A subset of the dialogue template and a corresponding dialogue example is shown in Figure 2.

4.2 Crowdsourced Data

Then we collected crowdsourced data via the Amazon MTurk platform. Participants were asked to answer two open-ended questions related to sleep issues and their impacts and check all applicable behavioural factors. Further, the participants are asked to paraphrase the specific sleep conditions (i.e., issues, impacts), if they have ever experienced them, and the selected behavioural factors. The former and the latter data are denoted as the answer data set and the paraphrase data set, respectively. The answer data set are further used to create user goals. Following the previous works (Schatzmann and Young, 2009; Wei et al., 2018; Xu et al., 2019), we create a user goal $G = (E, I)$ consisting of explicit information E, which is reported in the answers to the open-ended questions, and implicit information I, which is the answers to the behavioural factor that can be retrieved via probing questions. Table 1 summarises the size of each data set and the details of each data set are given in Appendix A.

Data set	Goal	Issue	Impact	Habit
Answer	3,015	3,015	3,015	7,961
Paraphrase	-	12,325	7,287	7,961

Table 1: Size of each data set.

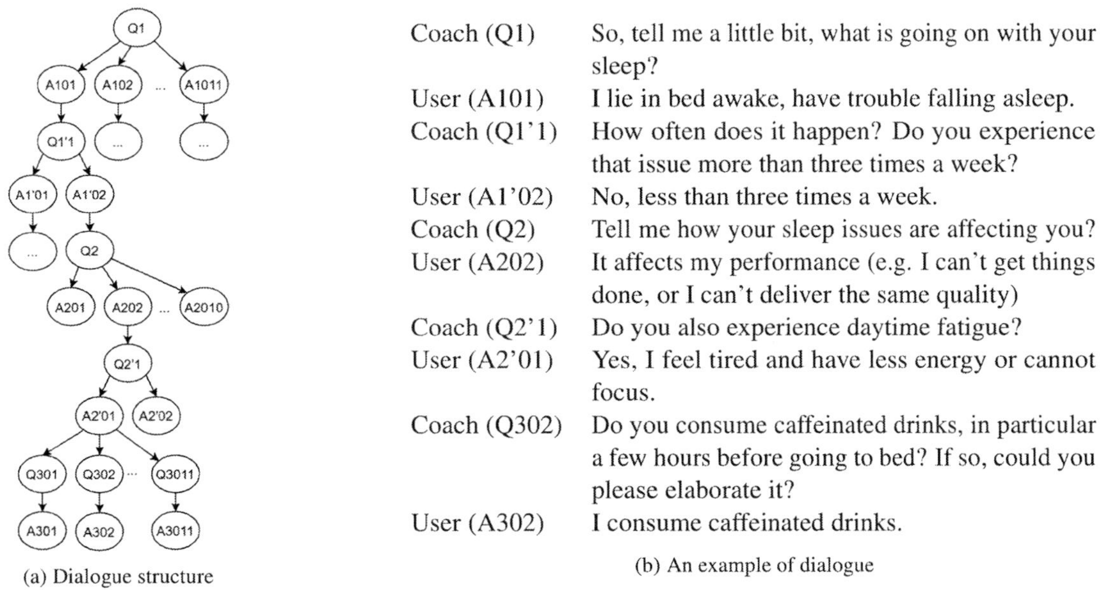

(a) Dialogue structure

Coach (Q1)	So, tell me a little bit, what is going on with your sleep?
User (A101)	I lie in bed awake, have trouble falling asleep.
Coach (Q1'1)	How often does it happen? Do you experience that issue more than three times a week?
User (A1'02)	No, less than three times a week.
Coach (Q2)	Tell me how your sleep issues are affecting you?
User (A202)	It affects my performance (e.g. I can't get things done, or I can't deliver the same quality)
Coach (Q2'1)	Do you also experience daytime fatigue?
User (A2'01)	Yes, I feel tired and have less energy or cannot focus.
Coach (Q302)	Do you consume caffeinated drinks, in particular a few hours before going to bed? If so, could you please elaborate it?
User (A302)	I consume caffeinated drinks.

(b) An example of dialogue

Figure 2: A subset of the dialogue template (left) and a corresponding dialogue example (right).

4.3 Dialogue Simulation

The collected crowdsourced data are further used to simulate dialogues. At the beginning of each dialogue, a user goal is sampled from the answer data set. Then a dialogue is simulated based on the dialogue template with a set of handcrafted rules and augmented by using the paraphrase data set. An example of a user goal and the simulated and augmented dialogues are shown in Appendix B.

4.4 Limitations and Future Study

In this paper, we show preliminary results of simulating dialogues based on the dialogue template and crowdsourced data. Our approach aims to augment the size of the simulated dialogue data set by replacing user answers with samples from the separate paraphrase data set. However, there are a few limitations that might be associated with the proposed method. More specifically, the following concerns should be addressed in a future study: First of all, the paraphrased sentences should be diverse and the simulated dialogues should cover all potential dialogue paths. To validate the quality, the paraphrased sentences and the simulated dialogues are required to be accessed by proper measures. Secondly, as Shi et al. (2019) has already pointed out, the RL agent may not generalise enough to real-world dialogues even though it works well with a user simulator. Therefore, there should be the additional step of on-line learning by interacting with real-world users (Shah et al., 2018) to mitigate this issue.

5 Conclusion

In this paper, we analyse recent studies on the development of a dialogue system in the healthcare domain and outline three building blocks, namely: i) privacy-preserving data collection; ii) medical knowledge-grounded dialogue management; and iii) human-centred evaluations. To this end, we propose a framework for developing a dialogue system and show preliminary results of simulated dialogue data generation by utilising expert knowledge and crowdsourcing. In the future study, we foresee working on implementing a user simulator that can interact with a reinforcement learning agent, accessing the quality of the simulated dialogues, and deploying the reinforcement learning agent to interact with both a user simulator and real-world users.

Acknowledgments

We thank anonymous reviewers for providing valuable feedback on this work. This project has received funding from the European Union's Horizon 2020 research and innovation programme under the Marie Skłodowska-Curie grant agreement No 766139. This article reflects only the author's view and the REA is not responsible for any use that may be made of the information it contains.

References

Daniel Adiwardana, Minh-Thang Luong, David R So, Jamie Hall, Noah Fiedel, Romal Thoppilan, Zi Yang, Apoorv Kulshreshtha, Gaurav Nemade, Yifeng Lu, et al. 2020. Towards a human-like open-domain chatbot. *arXiv preprint arXiv:2001.09977*.

Paweł Budzianowski and Ivan Vulić. 2019. Hello, it's gpt-2-how can i help you? towards the use of pre-trained language models for task-oriented dialogue systems. In *Proceedings of the 3rd Workshop on Neural Generation and Translation*, pages 15–22.

Jun Chen, Xiaoya Dai, Quan Yuan, Chao Lu, and Haifeng Huang. 2020. Towards interpretable clinical diagnosis with bayesian network ensembles stacked on entity-aware cnns. In *Proceedings of the 58th Annual Meeting of the Association for Computational Linguistics*, pages 3143–3153.

Layla El Asri, Jing He, and Kaheer Suleman. 2016. A sequence-to-sequence model for user simulation in spoken dialogue systems. *Interspeech 2016*, pages 1151–1155.

Shameek Ghosh, Sammi Bhatia, and Abhi Bhatia. 2018. Quro: facilitating user symptom check using a personalised chatbot-oriented dialogue system. *Stud Health Technol Inform*, 252:51–56.

Jiaqi Guan, Runzhe Li, Sheng Yu, and Xuegong Zhang. 2019. A method for generating synthetic electronic medical record text. *IEEE/ACM transactions on computational biology and bioinformatics*.

Amartya Hatua, Trung T Nguyen, and Andrew H Sung. 2019. Dialogue generation using self-attention generative adversarial network. In *2019 IEEE International Conference on Conversational Data & Knowledge Engineering (CDKE)*, pages 33–38. IEEE.

Hao-Cheng Kao, Kai-Fu Tang, and Edward Chang. 2018. Context-aware symptom checking for disease diagnosis using hierarchical reinforcement learning. In *Proceedings of the AAAI Conference on Artificial Intelligence*, volume 32.

Florian Kreyssig, Iñigo Casanueva, Paweł Budzianowski, and Milica Gasic. 2018. Neural user simulation for corpus-based policy optimisation of spoken dialogue systems. In *Proceedings of the 19th Annual SIGdial Meeting on Discourse and Dialogue*, pages 60–69.

Mike Lewis, Yinhan Liu, Naman Goyal, Marjan Ghazvininejad, Abdelrahman Mohamed, Omer Levy, Veselin Stoyanov, and Luke Zettlemoyer. 2020. Bart: Denoising sequence-to-sequence pretraining for natural language generation, translation, and comprehension. In *Proceedings of the 58th Annual Meeting of the Association for Computational Linguistics*, pages 7871–7880.

Margaret Li, Jason Weston, and Stephen Roller. 2019. Acute-eval: Improved dialogue evaluation with optimized questions and multi-turn comparisons. *arXiv preprint arXiv:1909.03087*.

Xiujun Li, Yun-Nung Chen, Lihong Li, Jianfeng Gao, and Asli Celikyilmaz. 2017. End-to-end task-completion neural dialogue systems. In *Proceedings of the Eighth International Joint Conference on Natural Language Processing (Volume 1: Long Papers)*, pages 733–743.

Anna Liednikova, Philippe Jolivet, Alexandre Durand-Salmon, and Claire Gardent. 2020. Learning healthbots from training data that was automatically created using paraphrase detection and expert knowledge. In *Proceedings of the 28th International Conference on Computational Linguistics*, pages 638–648.

Zhengyuan Liu, Hazel Lim, Nur Farah Ain Suhaimi, Shao Chuen Tong, Sharon Ong, Angela Ng, Sheldon Lee, Michael R Macdonald, Savitha Ramasamy, Pavitra Krishnaswamy, et al. 2019. Fast prototyping a dialogue comprehension system for nurse-patient conversations on symptom monitoring. In *Proceedings of the 2019 Conference of the North American Chapter of the Association for Computational Linguistics: Human Language Technologies, Volume 2 (Industry Papers)*, pages 24–31.

Stephane M Meystre, F Jeffrey Friedlin, Brett R South, Shuying Shen, and Matthew H Samore. 2010. Automatic de-identification of textual documents in the electronic health record: a review of recent research. *BMC medical research methodology*, 10(1):1–16.

Katherine Middleton, Mobasher Butt, Nils Hammerla, Steven Hamblin, Karan Mehta, and Ali Parsa. 2016. Sorting out symptoms: design and evaluation of the'babylon check'automated triage system. *arXiv preprint arXiv:1606.02041*.

Volodymyr Mnih, Koray Kavukcuoglu, David Silver, Andrei A Rusu, Joel Veness, Marc G Bellemare, Alex Graves, Martin Riedmiller, Andreas K Fidjeland, Georg Ostrovski, et al. 2015. Human-level control through deep reinforcement learning. *nature*, 518(7540):529–533.

Charles M Morin, Richard R Bootzin, Daniel J Buysse, Jack D Edinger, Colin A Espie, and Kenneth L Lichstein. 2006. Psychological and behavioral treatment of insomnia: update of the recent evidence (1998–2004). *Sleep*, 29(11):1398–1414.

Ishna Neamatullah, Margaret M Douglass, H Lehman Li-wei, Andrew Reisner, Mauricio Villarroel, William J Long, Peter Szolovits, George B Moody, Roger G Mark, and Gari D Clifford. 2008. Automated de-identification of free-text medical records. *BMC medical informatics and decision making*, 8(1):1–17.

Thomas Neubauer and Johannes Heurix. 2011. A methodology for the pseudonymization of medical data. *International journal of medical informatics*, 80(3):190–204.

Lin Ni, Chenhao Lu, Niu Liu, and Jiamou Liu. 2017. Mandy: Towards a smart primary care chatbot application. In *International symposium on knowledge and systems sciences*, pages 38–52. Springer.

Youcheng Pan, Qingcai Chen, Weihua Peng, Xiaolong Wang, Baotian Hu, Xin Liu, Junying Chen, and Wenxiu Zhou. 2020. Medwriter: Knowledge-aware medical text generation. In *Proceedings of the 28th International Conference on Computational Linguistics*, pages 2363–2368.

Alec Radford, Karthik Narasimhan, Tim Salimans, and Ilya Sutskever. 2018. Improving language understanding with unsupervised learning.

Alec Radford, Jeffrey Wu, Rewon Child, David Luan, Dario Amodei, and Ilya Sutskever. 2019. Language models are unsupervised multitask learners. *OpenAI blog*, 1(8):9.

Hannah Rashkin, Eric Michael Smith, Margaret Li, and Y-Lan Boureau. 2019. Towards empathetic open-domain conversation models: A new benchmark and dataset. In *Proceedings of the 57th Annual Meeting of the Association for Computational Linguistics*, pages 5370–5381.

Salman Razzaki, Adam Baker, Yura Perov, Katherine Middleton, Janie Baxter, Daniel Mullarkey, Davinder Sangar, Michael Taliercio, Mobasher Butt, Azeem Majeed, et al. 2018. A comparative study of artificial intelligence and human doctors for the purpose of triage and diagnosis. *arXiv preprint arXiv:1806.10698*.

Stephen Roller, Emily Dinan, Naman Goyal, Da Ju, Mary Williamson, Yinhan Liu, Jing Xu, Myle Ott, Kurt Shuster, Eric M Smith, et al. 2020. Recipes for building an open-domain chatbot. *arXiv preprint arXiv:2004.13637*.

Jost Schatzmann and Steve Young. 2009. The hidden agenda user simulation model. *IEEE transactions on audio, speech, and language processing*, 17(4):733–747.

Pararth Shah, Dilek Hakkani-Tur, Bing Liu, and Gokhan Tur. 2018. Bootstrapping a neural conversational agent with dialogue self-play, crowdsourcing and on-line reinforcement learning. In *Proceedings of the 2018 Conference of the North American Chapter of the Association for Computational Linguistics: Human Language Technologies, Volume 3 (Industry Papers)*, pages 41–51.

Emily Sheng, Kai-Wei Chang, Premkumar Natarajan, and Nanyun Peng. 2019. The woman worked as a babysitter: On biases in language generation. In *Proceedings of the 2019 Conference on Empirical Methods in Natural Language Processing and the*

9th International Joint Conference on Natural Language Processing (EMNLP-IJCNLP), pages 3407–3412, Hong Kong, China. Association for Computational Linguistics.

Weiyan Shi, Kun Qian, Xuewei Wang, and Zhou Yu. 2019. How to build user simulators to train rl-based dialog systems. In *Proceedings of the 2019 Conference on Empirical Methods in Natural Language Processing and the 9th International Joint Conference on Natural Language Processing (EMNLP-IJCNLP)*, pages 1990–2000.

Ashish Vaswani, Noam Shazeer, Niki Parmar, Jakob Uszkoreit, Llion Jones, Aidan N Gomez, Łukasz Kaiser, and Illia Polosukhin. 2017. Attention is all you need. *Advances in Neural Information Processing Systems*, 30:5998–6008.

Eric Wallace, Shi Feng, Nikhil Kandpal, Matt Gardner, and Sameer Singh. 2019. Universal adversarial triggers for attacking and analyzing NLP. In *Proceedings of the 2019 Conference on Empirical Methods in Natural Language Processing and the 9th International Joint Conference on Natural Language Processing (EMNLP-IJCNLP)*, pages 2153–2162, Hong Kong, China. Association for Computational Linguistics.

Zhongyu Wei, Qianlong Liu, Baolin Peng, Huaixiao Tou, Ting Chen, Xuan-Jing Huang, Kam-Fai Wong, and Xiang Dai. 2018. Task-oriented dialogue system for automatic diagnosis. In *Proceedings of the 56th Annual Meeting of the Association for Computational Linguistics (Volume 2: Short Papers)*, pages 201–207.

Thomas Wolf, Victor Sanh, Julien Chaumond, and Clement Delangue. 2019. Transfertransfo: A transfer learning approach for neural network based conversational agents. *arXiv preprint arXiv:1901.08149*.

Lin Xu, Qixian Zhou, Ke Gong, Xiaodan Liang, Jianheng Tang, and Liang Lin. 2019. End-to-end knowledge-routed relational dialogue system for automatic diagnosis. In *Proceedings of the AAAI Conference on Artificial Intelligence*, volume 33, pages 7346–7353.

Steve Young, Milica Gašić, Blaise Thomson, and Jason D Williams. 2013. Pomdp-based statistical spoken dialog systems: A review. *Proceedings of the IEEE*, 101(5):1160–1179.

Guangtao Zeng, Wenmian Yang, Zeqian Ju, Yue Yang, Sicheng Wang, Ruisi Zhang, Meng Zhou, Jiaqi Zeng, Xiangyu Dong, Ruoyu Zhang, et al. 2020. Meddialog: A large-scale medical dialogue dataset. In *Proceedings of the 2020 Conference on Empirical Methods in Natural Language Processing (EMNLP)*, pages 9241–9250.

Yizhe Zhang, Siqi Sun, Michel Galley, Yen-Chun Chen, Chris Brockett, Xiang Gao, Jianfeng Gao, Jingjing

Liu, and William B Dolan. 2020. Dialogpt: Large-scale generative pre-training for conversational response generation. In *Proceedings of the 58th Annual Meeting of the Association for Computational Linguistics: System Demonstrations*, pages 270–278.

A Crowdsourced Data

We collected two crowdsourced data sets for experiments: The answer data set contains user goals consisting of answers to the two open-ended questions (i.e., sleep issue and the impact of the issue) and one multiple-choice question (i.e., habits/lifestyles). The paraphrase data set contains paraphrased answers related to the sleep conditions (i.e., sleep issue and the impact of the issue) and the selected multiple-choice answers (i.e., habits/lifestyles). The collected data were annotated with class labels as shown in tables 2 to 4. Figure 3 shows label distributions of the collected data sets.

Class	Description
troubleFallingAsleep	Lie in bed awake
troubleStayingAsleep	Wake up frequently
staysUpLate	Stay up late
wakeUpTooEarly	Wake up too early
problemWakingUp	Trouble waking up
sleepsInLater	Sleep in late
snoringBothersMe	Snoring issue 1
snoringBothersOthers	Snoring issue 2
snoringStoppedBreathing	Breathing problem
otherIssue	Other issue
goodSleep	No issue

Table 2: Class labels for sleep issues.

Class	Description
energy	Feel tired or less energy
performance	Affect performance
embarrassedBySnoring	Snoring impact
dryMouth	Cause dry mouth
appearance	Look tired
stressMoodAnxiety	Bad mood
lessPatience	Become less patience
socialImpact	Affect social life
otherHealthImmunity	Affect health
noImpact	No impact

Table 3: Class labels for the impacts of sleep issues.

Class	Description
media	Engage in screen-time
caffeine	Consume caffeine
drinking	Consume drink
alcohol	Consume alcohol drinks
nicotine	Smoke
eating	Eat heavy meals
exercise	Work out/exercise
passivity	Physically not active
napping	Nap during the day
obligationDuties	Too many duties
stressMoodAnxiety	Experience stress

Table 4: Class labels for habits/lifestyles.

B User Goal and Simulated Dialogue

An example of a user goal is shown in Figure 4. To simulate a dialogue, we used the dialogue template with a set of handcrafted rules to select a coach's next question. Each question is followed by the answer by using the sampled user goal. If the question cannot be answered by the user goal, we randomly select an answer either *Yes* or *No*. The simulated dialogue is then paraphrased by replacing user answers with samples from the paraphrase data set. Table 5 illustrates the examples of a simulated dialogue and an augmented dialogue.

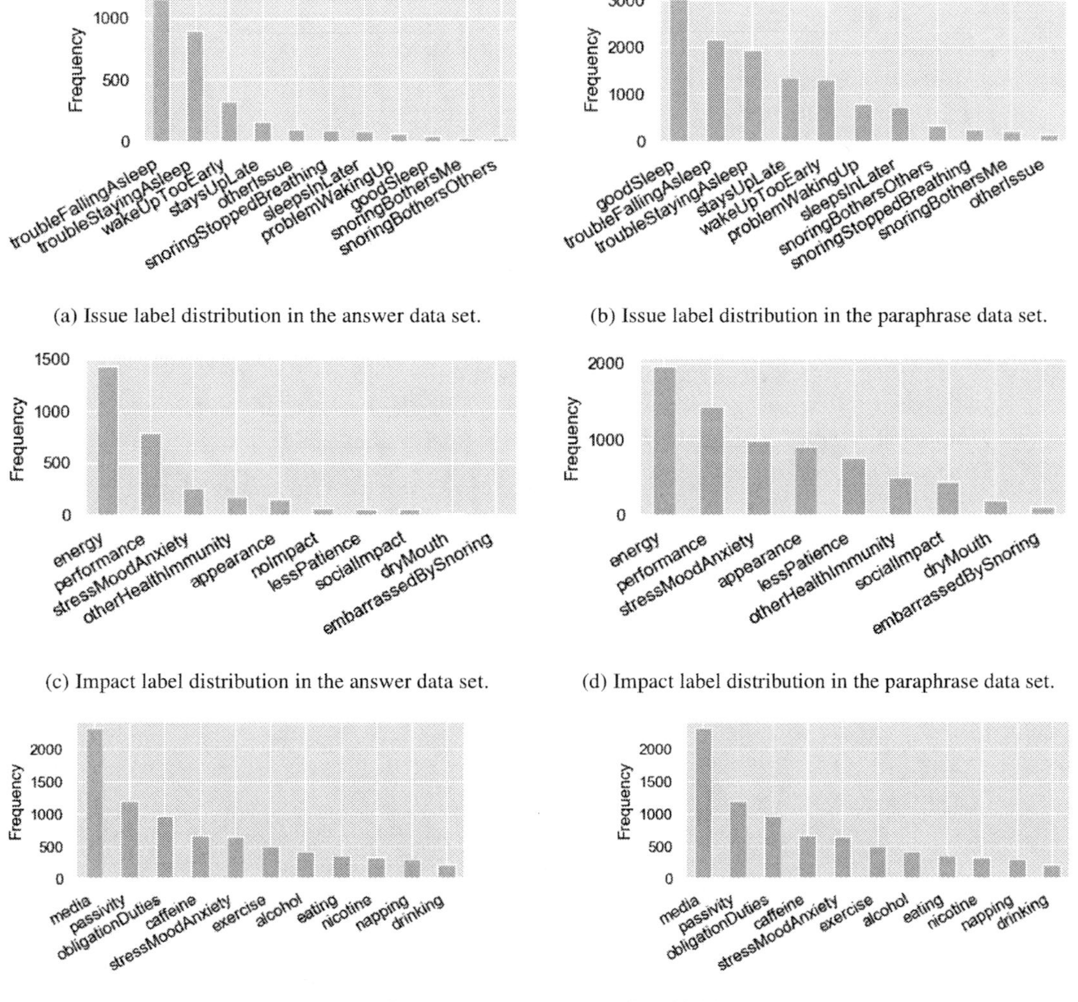

(a) Issue label distribution in the answer data set.

(b) Issue label distribution in the paraphrase data set.

(c) Impact label distribution in the answer data set.

(d) Impact label distribution in the paraphrase data set.

(e) Habit label distribution in the answer data set.

(f) Habit label distribution in the paraphrase data set.

Figure 3: Class label distributions of the collected data sets. Note that the answer data set and the paraphrase data set have identical habit class label distribution but the former contains binary values (i.e., True, False) and the latter contains free-text values (i.e., paraphrased sentences).

	A simulated dialogue
Coach	So, tell me a little bit, what is going on with your sleep?
User	*I just can't get to sleep.*
	<u>*I lie in bed awake, have trouble falling asleep.*</u>
	<u>*I think too much about work issues and need to stop doing that.*</u>
	<u>*I try to fall asleep, but I just lay there. The sleep doesn't come for me quickly and I have to wait and wait until my body finally falls asleep.*</u>
Coach	How often does it happen? Do you experience that issue more than three times a week?
User	Yes.
Coach	How long does your issue last in general? More than 30 minutes?
User	No.
Coach	Tell me how your sleep issues are affecting you?
User	*My exhaustion really affects my work. I'm not sharp like I used to. I feel tortured.*
	<u>*I do less because I'm exhausted.*</u>
	<u>*I need more time to get things done, and I don't have the creativity and energy that I would want to deliver top quality work.*</u>
	<u>*Because I have not received enough sleep I do not focus as well. This causes my performance to not be as well as it should.*</u>
Coach	Do you also experience daytime fatigue?
User	No
Coach	Do you experience stress or mood swings?
User	No
Coach	Do you engage with digital devices/screen, in particular, a few hours before going to bed?
User	Yes
	<u>*I'm around screens all the time and it affects my sleep.*</u>
	<u>*I end up being on my computer working all day and when I'm not working I'm watching TV or on my phone. I do these things immediately before going to bed and while in bed.*</u>
	<u>*Most of the time leading up to going to bed for us is watching TV. But really this is just about the only time I have to look through facebook, and emails on my phone too. So it's like I'm getting a double whammy of light from these devices.*</u>

Table 5: An example of a simulated dialogue based on the dialogue template with a sampled user goal and paraphrased sentences. Italic texts are the source texts extracted from the user goal and underlined italic texts are target sentences sampled from the paraphrased data set. Three randomly sampled paraphrased sentences per user answer are reported.

```
{
'explicit':{
    'main_issue': 'troubleFallingAsleep
        ',
    'main_issue_text': "I just can't
        get to sleep.",
    'main_impact': 'performance',
    'main_impact_text': "My exhaustion
        really affects my work. I'm not
         sharp like I used to. I feel
        tortured.",
},
'implicit': {
    'passivity': False,
    'alcohol': False,
    'nicotine': False,
    'caffeine': False,
    'media': True,
    'exercise': False,
    'drinking': False,
    'eating': False,
    'stressMoodAnxiety': False,
    'obligationDuties': False,
    'napping': False
}
}
```

Figure 4: An example of a user goal.

Joint Summarization-Entailment Optimization
for Consumer Health Question Understanding

Khalil Mrini[1], Franck Dernoncourt[2],
Walter Chang[2], Emilia Farcas[1], and Ndapa Nakashole[1]

[1]University of California, San Diego, La Jolla, CA 92093
{khalil, efarcas, nnakashole}@ucsd.edu
[2]Adobe Research, San Jose, CA 95110
{franck.dernoncourt, wachang}@adobe.com

Abstract

Understanding the intent of medical questions asked by patients, or Consumer Health Questions, is an essential skill for medical Conversational AI systems. We propose a novel data-augmented and simple joint learning approach combining question summarization and Recognizing Question Entailment (RQE) in the medical domain. Our data augmentation approach enables to use just one dataset for joint learning. We show improvements on both tasks across four biomedical datasets in accuracy (+8%), ROUGE-1 (+2.5%) and human evaluation scores. Human evaluation shows joint learning generates faithful and informative summaries. Finally, we release our code, the two question summarization datasets extracted from a large-scale medical dialogue dataset, as well as our augmented datasets[1].

1 Introduction

In order to answer questions, Conversational AI systems have to first understand the intent of questions (Chen et al., 2012; Cai et al., 2017). This is particularly important for medical conversational agents (Wu et al., 2020), as Consumer Health Questions (CHQ) are often long and contain peripheral information not needed to answer the question. Approaches to medical question understanding include query relaxation (Ben Abacha and Zweigenbaum, 2015; Lei et al., 2020), question entailment recognition (Ben Abacha and Demner-Fushman, 2016, 2019b; Agrawal et al., 2019) and summarization (Ben Abacha and Demner-Fushman, 2019a).

We approach the problem of medical question understanding using joint learning of medical question pairs in the two tasks of question summarization and Recognizing Question Entailment (RQE). Previous work on combining summarization and entailment uses at least two datasets – one for each

---

[1]https://github.com/KhalilMrini/
Medical-Question-Understanding

task. We start from the observation that, given a pair of questions A and B, where A is the longer question, A entails B if and only if B is a summary of A. Using this observation, we propose a data augmentation scheme to use a single dataset for joint learning, instead of two. Then, we propose a simple, simultaneous joint learning approach with fully shared model parameters.

Our findings show that joint learning performs significantly better than single-task training. Our joint learning approach brings about an 8% increase in accuracy in the RQE task compared to single-task training, and shows an average of 2.5% increase in ROUGE-1 F1 scores across three medical question summarization datasets. Additionally, we perform human evaluation and find our approach generates more informative question summaries. Our results suggest the RQE objective makes our summaries more similar in style to the CHQ. Finally, we release the two consumer health question summarization datasets we extracted from an existing large-scale medical dialogue dataset, our augmented datasets and our code.

2 Background and Related Work

2.1 Recognizing Question Entailment (RQE)

The task of RQE was introduced by Ben Abacha and Demner-Fushman (2016) in the context of medical question answering. It is closely related to the task of Recognizing Textual Entailment (RTE) (Dagan et al., 2005, 2013), and early definitions of question entailment (Groenendijk and Stokhof, 1984; Roberts, 1996). Ben Abacha and Demner-Fushman (2016) define RQE as follows: given a pair of questions A and B, question A entails question B if every answer to B is a correct answer to A, and answers A either partially or fully.

2.2 Transfer Learning for Medical QA

Language models that use multi-task learning and transfer learning have become ubiquitous in various

Proceedings of the Second Workshop on Natural Language Processing for Medical Conversations, pages 58–65
July 6, 2021. ©2021 Association for Computational Linguistics

NLP applications, including BioNLP. BERT (Devlin et al., 2019) has been fine-tuned using biomedical text from PubMed (Beltagy et al., 2019), PMC (Lee et al., 2020), and/or the MIMIC III dataset (Johnson et al., 2016; Huang et al., 2019; Alsentzer et al., 2019). In this paper, we use pre-trained BART models (Lewis et al., 2019).

Transfer learning was a popular approach at the 2019 MEDIQA shared task (Ben Abacha et al., 2019) on medical NLI, RQE and QA. The question answering task involved re-ranking answers, not generating them (Demner-Fushman et al., 2020). For the RQE task, the best-performing model (Zhu et al., 2019) uses transfer learning on NLI and ensemble methods.

In contemporaneous work of ours (Mrini et al., 2021), we participate in the question summarization task of the 2021 MEDIQA shared task (Ben Abacha et al., 2021). We show that transfer learning using medical RQE can improve performance on medical question summarization.

2.3 Summarization and Entailment

There is a growing body of work combining summarization and entailment (Lloret et al., 2008; Mehdad et al., 2013; Gupta et al., 2014).

Falke et al. (2019) use textual entailment predictions to detect factual errors in abstractive summaries generated by state-of-the-art models. Pasunuru and Bansal (2018) propose an entailment reward for their abstractive summarizer, where the entailment score is obtained from a pre-trained and frozen natural language inference model.

Pasunuru et al. (2017) propose an LSTM encoder-decoder model that incorporates entailment generation and abstractive summarization. They use separate natural language inference and summarization datasets, and train by optimizing the two objectives alternatively. Guo et al. (2018) build upon the work of Pasunuru et al. (2017), and add question generation as an auxiliary task.

Li et al. (2018) propose an encoder-decoder summarization model, with an entailment-aware encoder with a separate classification module, and an entailment-rewarded decoder. They follow closely the multi-task setting of Pasunuru et al. (2017).

3 Joint Learning for Consumer Health Question Understanding

We consider the joint learning of medical question summarization and Recognizing Question Entailment (RQE). In both tasks, a question pair includes a first medical question, written in an informal style by a patient – thus called a Consumer Health Question (CHQ). The second medical question is shorter, and often written in a formal style by medical experts: it is a Frequently Asked Question (FAQ). The inspiration for our joint learning scheme stems from the observation that a CHQ entails an FAQ, if and only if the FAQ is a summary of the CHQ.

Our data-augmented joint learning approach to consumer health question understanding has two main components. First, we use our equivalence observation to propose a scheme for data augmentation. Second, we show our joint learning model architecture and learning objective.

3.1 Data Augmentation

Instead of using separate datasets as in previous work, we propose to augment datasets to train jointly, such that we have the same amount of summarization and RQE pairs.

For summarization datasets, we create equivalent RQE pairs. For each existing summarization pair, we first choose with equal probability whether the equivalent RQE pair is labeled as entailment or not. If it is an entailment case, we create an RQE pair identical to the summarization pair. If it is not an entailment case, the CHQ of the RQE pair is identical to the CHQ of the summarization pair, and the FAQ of the RQE pair is a different, randomly selected from the FAQs of the same dataset split.

Inversely, for the RQE dataset, we create equivalent summarization pairs. For each existing RQE pair, we consider two cases. If the RQE pair is labeled as entailment, we create an identical summarization pair. If the RQE pair is labeled as not entailment, we create a summarization pair that is identical to a randomly selected entailment-labeled RQE pair from the same dataset split.

3.2 Joint Model

We adopt the architecture of BART Large (Lewis et al., 2019), a model that set a new state of the art in XSum (Narayan et al., 2018) and CNN-Dailymail (Hermann et al., 2015), two popular abstractive summarization benchmark datasets.

BART is an encoder-decoder seq2seq model, that can train generation as well as classification tasks, such as RQE. BART trains for abstractive summarization by feeding the source text (CHQ) to the encoder, and the negative log-likelihood loss is computed between the decoder output and the

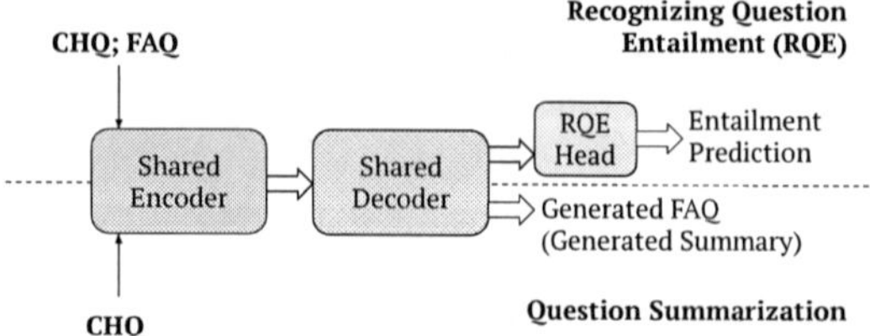

Figure 1: An example medical question pair. The first question is a Consumer Health Question (CHQ) and the second question is a Frequently Asked Question (FAQ). We use BART (Lewis et al., 2019) to jointly train question summarization (bottom) and RQE (top). We show how BART takes input differently for each task.

reference summary (FAQ). BART trains for classification by feeding the full input to the encoder – in the case of RQE, the full input is the concatenation of the CHQ and FAQ. An added classification head attached to the last decoder output then generates a prediction. We compute the binary cross-entropy loss based on the classification head's prediction and the RQE label. We show an overview of our joint training in Figure 1.

We propose to optimize a single loss function that is the sum of the objectives of both tasks. At each training step, we have a summarization question pair that is used for the negative log-likelihood loss, and an RQE question pair that is used for the Binary Cross-Entropy (BCE) loss. Given a CHQ embedding x, the corresponding FAQ embedding y, and the entailment label $l_{entail} \in \{0, 1\}$, we optimize the following loss function:

$$\mathcal{L}_{joint} = -\log p(y|x; \theta) + \text{BCE}(x, y, l_{entail}; \theta) \tag{1}$$

For RQE, we consider two loss alternatives, in which we create summarization pairs that are identical to the RQE pairs, regardless of entailment. In the first alternative we simply remove the negative log-likelihood loss for pairs labeled as not entailment. In the second alternative, we flip the negative log-likelihood loss for pairs labeled as not entailment, such that we try to maximize the summarization loss instead of minimizing it.

Dataset	Train	Dev	Test
MeQSum	400	100	500
HealthCareMagic	181,122	22,641	22,642
iCliniq	24,851	3,105	3,106
MEDIQA RQE	8,588	302	230

Table 1: Statistics of the medical dataset splits.

4 Experiment Setup

4.1 Datasets

We consider three medical question summarization datasets and one medical RQE dataset, all in English. Table 1 shows dataset statistics.

(1) MeQSum (Ben Abacha and Demner-Fushman, 2019a) is a medical question summarization dataset released by the U.S. National Institutes of Health (NIH). It contains 1,000 consumer health questions summarized into FAQ-style single-sentence questions by medical experts. The authors used the first 500 datapoints as training and the last 500 as testing. We use a randomly selected 100 datapoints from the training set as our dev set.

We extract the **(2) HealthCareMagic and (3) iCliniq** question summarization datasets from Med-Dialog (Zeng et al., 2020), a large-scale medical dialogue dataset collected from two online healthcare service platforms: `HealthCareMagic.com` and `iCliniq.com`.

These two datasets include first a one-sentence question describing the medical condition of the patient, followed by two long utterances: one from the patient that includes a description of the problem and a question, and then one from the doctor that includes the response. To form medical question summarization datasets, we consider the single-sentence descriptions as summaries of the patient utterances. HealthCareMagic's summaries are more abstractive and are written in a formal style, unlike iCliniq's patient-written summaries. We create a 80/10/10 split for train/dev/test sets.

(4) MEDIQA RQE is the RQE dataset of the 2019 MEDIQA shared task (Ben Abacha et al., 2019). The test set comprises manually written question pairs, whereas the train and dev sets (Ben Abacha and Demner-Fushman, 2016) are automatically collected. This difference explains the higher dev set results in Ben Abacha et al. (2019). Similarly to MeQSum, the question pairs match a longer CHQ received by the US National Library of Medicine (NLM) and a FAQ from the NIH.

Dataset	MeQSum			HealthCareMagic			iCliniq		
Metric	R1	R2	RL	R1	R2	RL	R1	R2	RL
Seq2seq Attentional Model (Nallapati et al., 2016)	24.8	13.8	24.3	-	-	-	-	-	-
Pointer-Generator Networks (PG) (See et al., 2017)	35.8	20.2	34.8	-	-	-	-	-	-
PG + Data Augmentation (Ben Abacha and Demner-Fushman, 2019a)	44.2	27.6	42.8	-	-	-	-	-	-
PG + Coverage Loss (See et al., 2017)	39.6	23.1	38.5	-	-	-	-	-	-
PG + Coverage Loss + Data Augmentation (Ben Abacha and Demner-Fushman, 2019a)	41.8	24.8	40.5	-	-	-	-	-	-
BART (Lewis et al., 2019)	45.7	26.8	40.8	**44.5**	**22.3**	**39.7**	48.7	28.0	43.5
BART + Data-Augmented Joint Learning	**48.5**	**29.7**	**44.9**	42.1	20.7	37.9	**53.5**	**36.5**	**48.6**

Table 2: Results on the test set comparing BART with and without joint learning of question summarization. The R1, R2 and RL metrics refer to the F1 scores of ROUGE-1, ROUGE-2 and ROUGE-L (Lin, 2004).

4.2 Setup

All of our models use the BART large architecture, with different pre-trained models for transfer learning. For the question summarization experiments, we use the BART Large model pre-trained on the XSum dataset (Narayan et al., 2018). For the RQE experiments, we pre-train a BART Large model on the RTE dataset (Dagan et al., 2005; Haim et al., 2006; Giampiccolo et al., 2007; Bentivogli et al., 2009) from the GLUE benchmark (Wang et al., 2018), and re-use the same classification head for RQE.

4.3 Training Settings

We train for 100 epochs for the MeQSum dataset, and for 10 epochs for all other datasets. We report ROUGE F1 scores for the question summarization datasets, and accuracy for the RQE dataset, as it is a binary classification task with two labels: entailment and not entailment.

For the question summarization datasets, the negative log likelihood on the dev set is used to select the best model. For the RQE dataset, the RQE accuracy on the dev set is the metric used to select the best model.

For single-task training, we use binary cross entropy for RQE, and negative log-likelihood for question summarization.

The learning rate for RQE experiments is 10^{-5} and for the question summarization experiments, it is $3 * 10^{-5}$. We use an Adam optimizer where the betas are 0.9 and 0.999 for summarization, and 0.9 and 0.98 for RQE. In all experiments, the Adam epsilon is 10^{-8}, and the dropout is 0.1.

4.4 Inference

At test time, we evaluate each task completely separately. For RQE, we feed the concatenation of the CHQ and FAQ as input to the model. For question summarization, we only feed the CHQ as input to the model. This way, we ensure that the model never sees the reference FAQs when being evaluated for question summarization.

5 Results and Discussion

5.1 Summarization Results

In their introduction of MeQSum, Ben Abacha and Demner-Fushman (2019a) show results with seq2seq models and pointer-generated networks. They additionally propose to augment MeQSum using semantically selected relevant pairs from the Quora Question Pairs dataset (Iyer et al., 2017). We report these baselines as well as our BART baseline results.

We show our summarization results in Table 2. On MeQSum and iCliniq, our joint learning objective achieves increases between 3 and 8 points across all three metrics – a significant improvement despite MeQSum being extremely low-resource. On the more abstractive and larger HealthCareMagic dataset, there is a decrease of 2 points compared to the BART baseline.

5.2 Human Evaluation

Given that ROUGE is notoriously unreliable, we hire 2 volunteer annotators, and we pick 40 generated summaries from each model in each summarization dataset, resulting in 240 generated summaries (FAQs). We collect 960 evaluations using best-worst scaling. The annotators could also choose to judge both generated FAQs as equal with regards to the given criteria. We show the annotators the generated FAQs in a random order, so that they do not know which model generated which FAQ. We evaluate the generated summaries on 4 criteria:

Datasets	Fluency	Coherence	Informative	Correct
MeQSum	+21.25%	+12.50%	+5.00%	−1.25%
HealthCareMagic	+3.75%	+8.75%	+11.25%	+2.50%
iCliniq	0%	−1.25%	−2.50%	0%

Table 3: Human Evaluation results on 120 samples from the question summarization datasets. The percentages indicate the added value of our joint learning.

Loss Function	Accuracy
Joint Learning	**78.1%**
Removing NLL if not entailment	73.1%
Maximizing NLL if not entailment	72.8%

Table 4: RQE accuracy results on the dev set of our joint loss compared to the two loss alternatives. NLL is Negative Log-Likelihood, the summarization loss.

- Fluency: which generated FAQ is more grammatically correct, and easier to read and to understand?

- Coherence: which generated FAQ is better structured and more organized?

- Informativeness: which generated FAQ captures the most out of the concern of the patient who wrote the CHQ?

- Correctness: which generated FAQ is more factually correct given the CHQ?

Our human evaluation results are in Table 3. Scores are generally in favor of our approach in MeQSum and HealthCareMagic. There is a high increase in informativeness for HealthCareMagic, and the results for iCliniq show that our approach gives summaries of roughly similar quality as the BART baseline. The ROUGE score increases in the extractive iCliniq and decreases in the abstractive HealthCareMagic indicate that our approach's summaries are more faithful to patient writing styles, suggesting a stronger influence from entailment.

5.3 RQE Results

We compare the joint loss function of equation 1 with the two loss alternatives in section 3.2. We show the results on the dev set in Table 4. Our

Method	Accuracy
BART (Lewis et al., 2019)	52.1%
Feature-based SVM (Ben Abacha and Demner-Fushman, 2016)	54.1%
BART + Data-Augmented Joint Learning	**60.0%**

Table 5: Accuracy results on MEDIQA RQE test set.

joint loss function fares the best, exceeding the alternatives by 5%. The results suggest that optimizing RQE jointly with question summarization does help improve performance on the RQE side as well. The difference with the alternative where we remove NLL for not-entailment pairs shows that optimizing our joint learning objective is more efficient than alternating single-task objectives.

We show our RQE results in Table 5. We see an 8% increase on the test set compared to optimizing only on the RQE objective. Our findings show that joint learning helps both tasks equally.

6 Conclusions

We propose a novel data-augmented joint learning approach for the tasks of RQE and question summarization. Our data augmentation method extends a dataset such that it can be used for both tasks. Our results show improvements in both tasks, across three question summarization datasets (+2.5% in ROUGE-1 F1) and one RQE dataset (+8% accuracy). We perform a human evaluation for our generated summaries: we find that our approach generates more informative summaries for formally written FAQs, and summaries that are faithful to patient writing styles in the more extractive iCliniq dataset. Finally, we make our datasets, code and training details publicly available.

Acknowledgments

We gratefully acknowledge the award from NIH/NIA grant R56AG067393. Khalil Mrini is additionally supported by unrestricted gifts from Adobe Research. We thank Naba Rizvi for the annotation work, and the anonymous reviewers for their feedback.

References

Anumeha Agrawal, Rosa Anil George, Selvan Suntiha Ravi, Sowmya Kamath, and Anand Kumar. 2019. Ars_nitk at MEDIQA 2019: analysing various methods for natural language inference, recognising question entailment and medical question answering system. In *Proceedings of the 18th BioNLP Workshop and Shared Task*, pages 533–540.

Emily Alsentzer, John Murphy, William Boag, Wei-Hung Weng, Di Jindi, Tristan Naumann, and Matthew McDermott. 2019. Publicly available clinical BERT embeddings. In *Proceedings of the 2nd Clinical Natural Language Processing Workshop*, pages 72–78.

Iz Beltagy, Kyle Lo, and Arman Cohan. 2019. SciB-ERT: A pretrained language model for scientific text. In *Proceedings of the 2019 Conference on Empirical Methods in Natural Language Processing and the 9th International Joint Conference on Natural Language Processing (EMNLP-IJCNLP)*, pages 3606–3611.

Asma Ben Abacha and Dina Demner-Fushman. 2016. Recognizing question entailment for medical question answering. In *AMIA Annual Symposium Proceedings*, volume 2016, page 310. American Medical Informatics Association.

Asma Ben Abacha and Dina Demner-Fushman. 2019a. On the summarization of consumer health questions. In *Proceedings of the 57th Annual Meeting of the Association for Computational Linguistics*, pages 2228–2234.

Asma Ben Abacha and Dina Demner-Fushman. 2019b. A question-entailment approach to question answering. *BMC bioinformatics*, 20(1):511.

Asma Ben Abacha, Yassine Mrabet, Yuhao Zhang, Chaitanya Shivade, Curtis Langlotz, and Dina Demner-Fushman. 2021. Overview of the mediqa 2021 shared task on summarization in the medical domain. In *Proceedings of the 20th SIG-BioMed Workshop on Biomedical Language Processing, NAACL-BioNLP 2021*. Association for Computational Linguistics.

Asma Ben Abacha, Chaitanya Shivade, and Dina Demner-Fushman. 2019. Overview of the MEDIQA 2019 shared task on textual inference, question entailment and question answering. In *Proceedings of the 18th BioNLP Workshop and Shared Task*, pages 370–379.

Asma Ben Abacha and Pierre Zweigenbaum. 2015. MEANS: A medical question-answering system combining NLP techniques and semantic web technologies. *Information processing & management*, 51(5):570–594.

Luisa Bentivogli, Peter Clark, Ido Dagan, and Danilo Giampiccolo. 2009. The fifth PASCAL recognizing textual entailment challenge. In *TAC*.

Ruichu Cai, Binjun Zhu, Lei Ji, Tianyong Hao, Jun Yan, and Wenyin Liu. 2017. A CNN-LSTM attention approach to understanding user query intent from online health communities. In *2017 IEEE International Conference on Data Mining Workshops (ICDMW)*, pages 430–437. IEEE.

Long Chen, Dell Zhang, and Levene Mark. 2012. Understanding user intent in community question answering. In *Proceedings of the 21st international conference on world wide web*, pages 823–828.

Ido Dagan, Oren Glickman, and Bernardo Magnini. 2005. The PASCAL recognising textual entailment challenge. In *Machine Learning Challenges Workshop*, pages 177–190. Springer.

Ido Dagan, Dan Roth, Mark Sammons, and Fabio Massimo Zanzotto. 2013. Recognizing textual entailment: Models and applications. *Synthesis Lectures on Human Language Technologies*, 6(4):1–220.

Dina Demner-Fushman, Yassine Mrabet, and Asma Ben Abacha. 2020. Consumer health information and question answering: helping consumers find answers to their health-related information needs. *Journal of the American Medical Informatics Association*, 27(2):194–201.

Jacob Devlin, Ming-Wei Chang, Kenton Lee, and Kristina Toutanova. 2019. BERT: Pre-training of deep bidirectional transformers for language understanding. In *Proceedings of the 2019 Conference of the North American Chapter of the Association for Computational Linguistics: Human Language Technologies, Volume 1 (Long and Short Papers)*, pages 4171–4186.

Tobias Falke, Leonardo FR Ribeiro, Prasetya Ajie Utama, Ido Dagan, and Iryna Gurevych. 2019. Ranking generated summaries by correctness: An interesting but challenging application for natural language inference. In *Proceedings of the 57th Annual Meeting of the Association for Computational Linguistics*, pages 2214–2220.

Danilo Giampiccolo, Bernardo Magnini, Ido Dagan, and Bill Dolan. 2007. The third PASCAL recognizing textual entailment challenge. In *Proceedings of the ACL-PASCAL workshop on textual entailment and paraphrasing*, pages 1–9.

Jeroen Antonius Gerardus Groenendijk and Martin Johan Bastiaan Stokhof. 1984. *Studies on the Semantics of Questions and the Pragmatics of Answers*. Ph.D. thesis, Univ. Amsterdam.

Han Guo, Ramakanth Pasunuru, and Mohit Bansal. 2018. Soft layer-specific multi-task summarization with entailment and question generation. In *Proceedings of the 56th Annual Meeting of the Association for Computational Linguistics (Volume 1: Long Papers)*, pages 687–697.

Anand Gupta, Manpreet Kaur, Shachar Mirkin, Adarsh Singh, and Aseem Goyal. 2014. Text summarization through entailment-based minimum vertex cover. In *Proceedings of the Third Joint Conference on Lexical and Computational Semantics (* SEM 2014)*, pages 75–80.

R Bar Haim, Ido Dagan, Bill Dolan, Lisa Ferro, Danilo Giampiccolo, Bernardo Magnini, and Idan Szpektor. 2006. The second PASCAL recognising textual entailment challenge. In *Proceedings of the Second PASCAL Challenges Workshop on Recognising Textual Entailment*.

Karl Moritz Hermann, Tomás Kociskỳ, Edward Grefenstette, Lasse Espeholt, Will Kay, Mustafa Suleyman, and Phil Blunsom. 2015. Teaching machines to read and comprehend. In *NIPS*.

Kexin Huang, Jaan Altosaar, and Rajesh Ranganath. 2019. ClinicalBERT: Modeling clinical notes and predicting hospital readmission. *arXiv preprint arXiv:1904.05342*.

Shankar Iyer, Nikhil Dandekar, and Kornél Csernai. 2017. First Quora dataset release: Question pairs.

Alistair EW Johnson, Tom J Pollard, Lu Shen, H Lehman Li-Wei, Mengling Feng, Mohammad Ghassemi, Benjamin Moody, Peter Szolovits, Leo Anthony Celi, and Roger G Mark. 2016. MIMIC-iii, a freely accessible critical care database. *Scientific data*, 3(1):1–9.

Jinhyuk Lee, Wonjin Yoon, Sungdong Kim, Donghyeon Kim, Sunkyu Kim, Chan Ho So, and Jaewoo Kang. 2020. BioBERT: a pretrained biomedical language representation model for biomedical text mining. *Bioinformatics*, 36(4):1234–1240.

Chuan Lei, Vasilis Efthymiou, Rebecca Geis, and Fatma Ozcan. 2020. Expanding query answers on medical knowledge bases. In *EDBT*, pages 567–578.

Mike Lewis, Yinhan Liu, Naman Goyal, Marjan Ghazvininejad, Abdelrahman Mohamed, Omer Levy, Ves Stoyanov, and Luke Zettlemoyer. 2019. BART: Denoising sequence-to-sequence pre-training for natural language generation, translation, and comprehension. *arXiv preprint arXiv:1910.13461*.

Haoran Li, Junnan Zhu, Jiajun Zhang, and Chengqing Zong. 2018. Ensure the correctness of the summary: Incorporate entailment knowledge into abstractive sentence summarization. In *Proceedings of the 27th International Conference on Computational Linguistics*, pages 1430–1441.

Chin-Yew Lin. 2004. ROUGE: A package for automatic evaluation of summaries. In *Text summarization branches out*, pages 74–81.

Elena Lloret, Oscar Ferrández, Rafael Munoz, and Manuel Palomar. 2008. A text summarization approach under the influence of textual entailment. In *NLPCS*, pages 22–31.

Yashar Mehdad, Giuseppe Carenini, Frank Tompa, and Raymond Ng. 2013. Abstractive meeting summarization with entailment and fusion. In *Proceedings of the 14th European Workshop on Natural Language Generation*, pages 136–146.

Khalil Mrini, Franck Dernoncourt, Seunghyun Yoon, Trung Bui, Walter Chang, Emilia Farcas, and Ndapa Nakashole. 2021. UCSD-Adobe at MEDIQA 2021: Transfer learning and answer sentence selection for medical summarization. In *Proceedings of the 20th SIGBioMed Workshop on Biomedical Language Processing, NAACL-BioNLP 2021*. Association for Computational Linguistics.

Ramesh Nallapati, Bowen Zhou, Cicero dos Santos, Çağlar Guİ‡lçehre, and Bing Xiang. 2016. Abstractive text summarization using sequence-to-sequence RNNs and beyond. In *Proceedings of The 20th SIGNLL Conference on Computational Natural Language Learning*, pages 280–290, Berlin, Germany. Association for Computational Linguistics.

Shashi Narayan, Shay B. Cohen, and Mirella Lapata. 2018. Don't give me the details, just the summary! Topic-aware convolutional neural networks for extreme summarization. In *Proceedings of the 2018 Conference on Empirical Methods in Natural Language Processing*, Brussels, Belgium.

Ramakanth Pasunuru and Mohit Bansal. 2018. Multi-reward reinforced summarization with saliency and entailment. In *Proceedings of the 2018 Conference of the North American Chapter of the Association for Computational Linguistics: Human Language Technologies, Volume 2 (Short Papers)*, pages 646–653.

Ramakanth Pasunuru, Han Guo, and Mohit Bansal. 2017. Towards improving abstractive summarization via entailment generation. In *Proceedings of the Workshop on New Frontiers in Summarization*, pages 27–32.

Craige Roberts. 1996. Information structure: Towards an integrated formal theory of pragmatics. *Semantics and Pragmatics*, 5:6–1.

Abigail See, Peter J Liu, and Christopher D Manning. 2017. Get to the point: Summarization with pointer-generator networks. In *Proceedings of the 55th Annual Meeting of the Association for Computational Linguistics (Volume 1: Long Papers)*, pages 1073–1083.

Alex Wang, Amanpreet Singh, Julian Michael, Felix Hill, Omer Levy, and Samuel Bowman. 2018. GLUE: A multi-task benchmark and analysis platform for natural language understanding. In *Proceedings of the 2018 EMNLP Workshop BlackboxNLP: Analyzing and Interpreting Neural Networks for NLP*, pages 353–355.

Chaochen Wu, Guan Luo, Chao Guo, Yin Ren, Anni Zheng, and Cheng Yang. 2020. An attention-based multi-task model for named entity recognition and intent analysis of chinese online medical questions. *Journal of Biomedical Informatics*, 108:103511.

Guangtao Zeng, Wenmian Yang, Zeqian Ju, Yue Yang, Sicheng Wang, Ruisi Zhang, Meng Zhou, Jiaqi Zeng, Xiangyu Dong, Ruoyu Zhang, Hongchao Fang, Penghui Zhu, Shu Chen, and Pengtao Xie. 2020. MedDialog: Large-scale medical dialogue datasets. In *Proceedings of the 2020 Conference on Empirical Methods in Natural Language Processing (EMNLP)*, pages 9241–9250, Online. Association for Computational Linguistics.

Wei Zhu, Xiaofeng Zhou, Keqiang Wang, Xun Luo, Xiepeng Li, Yuan Ni, and Guotong Xie. 2019. Panlp at MEDIQA 2019: Pre-trained language models, transfer learning and knowledge distillation. In *Proceedings of the 18th BioNLP Workshop and Shared Task*, pages 380–388.

Medically Aware GPT-3 as a Data Generator for Medical Dialogue Summarization

Bharath Chintagunta
jaic4@vt.edu

Namit Katariya
namit@curai.com

Xavier Amatriain
xavier@curai.com

Anitha Kannan
anitha@curai.com

Abstract

In medical dialogue summarization, summaries must be coherent and must capture all the medically relevant information in the dialogue. However, learning effective models for summarization require large amounts of labeled data which is especially hard to obtain. We present an algorithm to create synthetic training data with an explicit focus on capturing medically relevant information. We utilize GPT-3 as the backbone of our algorithm and scale 210 human labeled examples to yield results comparable to using 6400 human labeled examples ($\sim$30x) leveraging low-shot learning and an ensemble method. In detailed experiments, we show that this approach produces high quality training data that can further be combined with human labeled data to get summaries that are strongly preferable to those produced by models trained on human data alone both in terms of medical accuracy and coherency.

1 Introduction

With increasing usage of telehealth platforms (Mann et al., 2020), large scale ecosystems of providers and patients have become apparent. This has exacerbated the need for comprehensive visit summaries of the medical dialogues by the attending practitioner in order to facilitate accurate hand-offs to other care providers or as a means of recording the interaction. However, having providers write summaries after each encounter is not only time consuming but also costly, limiting the scalability of telehealth platforms (Shanafelt et al., 2016)

In these settings, an automated summarizer that can assist the practitioners can be extremely valuable. However, an important challenge of end-to-end medical dialogue summarization is the lack of large scale annotated datasets. Annotation of medical dialogues is expensive and slow because they need to be curated by trained experts. This is further compounded by the fact that labeled data may not be publicly shared because of patient privacy concerns and HIPAA regulations.

Recent approaches to summarization (Qi et al., 2020; Zhang et al., 2019) use transfer learning where a pre-trained model (e.g. through self supervision of learning a language model) is fine tuned with a labeled dataset. However, fine-tuning still requires hundreds to thousands of labeled examples to obtain reasonable performance. Methods such as (Joshi et al., 2020) aim to partially overcome these issues through modeling strategies that directly learn important inductive biases from smaller amounts of data. In addition, (Joshi et al., 2020) also handled data sparsity by leveraging a key insight of sequential nature of information flow in a medical dialogue: global summary of the dialogue can be composed from local dialogue turns (snippets). This enables collecting training data for snippets as opposed to the full conversation - an insight, we use in our paper as well.

Recently, OpenAI developed GPT-3, a neural language model that is capable of natural language generation and completion of tasks like classification, question-answering, and summarization (Brown et al., 2020). The focus of that work is to enable task-agnostic and zero-shot or low-shot performance as opposed to a pre-trained model that needs to be fine-tuned separately on every downstream task. In this paper, we investigate the following question: *How can a low-shot learner such as GPT-3 be leveraged to scale training data for medical dialogue summarization models?* In answering this question within the context of GPT-3 as a black box proprietary API[1], we took into account multiple considerations:

- *Medical Correctness (Joshi et al., 2020)*: Medical summarization warrants high recall and therefore the summarizer should be good at (1) capturing all the medical information (med-

---

[1] https://beta.openai.com/

Proceedings of the Second Workshop on Natural Language Processing for Medical Conversations, pages 66–76
July 6, 2021. ©2021 Association for Computational Linguistics

ications, symptoms, etc.) discussed in the dialogue and (2) discern all the affirmatives and negatives on medical conditions correctly (e.g. no allergies, having a cough for 2 days).

- *Privacy Concerns*: At inference time, an API call to external services such GPT-3 may not always be possible due to HIPAA and privacy concerns.

- *Practitioner in the loop*: The technique needs to be easily amenable to a feedback loop that allows for leveraging manually curated human labels. This feedback loop is extremely important because the diversity and the long tail of data distribution in medical dialogue means that there can be parts of the summary that need to be edited by practitioners for medical correctness and completeness. Note that these edits can be used as additional data for improving the underlying model.

Taking into account these considerations, this paper makes the following contributions (Figure 1 for a quick overview):

- We introduce a medically-aware GPT-3 data labeler, GPT-3-ENS , that combines medical knowledge and an ensemble of GPT-3 for the purpose of medical dialogue summarization.

- We introduce the idea of using GPT-3-ENS as a dataset generator to facilitate learning an in-house summarization model. Our experiments show that we can obtain the same performance as that of human labeled dataset with 30x smaller amount of human labeled data. With only 210 expert curated summaries and GPT-3 as a labeled data simulator, we can mimic the performance of a summarization model trained on 6400 expert curated summaries.

- By combining generated datasets from GPT-3-ENS with a human labeled dataset, we show that we can obtain better performance than models trained on either one of the data sources.

The rest of the paper is structured as follows: § 2 discusses related work, § 3 explores whether GPT-3 can be used directly for medical summarization, § 4 introduces our approach, § 5 and § 6 describe

our datasets and metrics respectively while § 7 illustrates our experiments. We end the paper with § 8 discussing our conclusions and future work.

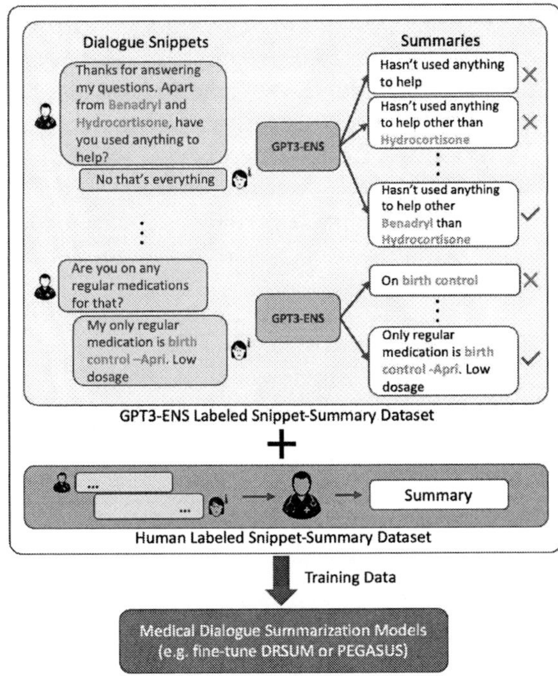

Figure 1: Overview of our proposed approach: we train models on a mix of GPT-3-ENS synthesized and human labeled data to get performance better than models trained on either of the sources

2   Related work

Summarization Emergence of sequence to sequence models and attention mechanisms (Sutskever et al., 2014) has led to rapid progress on extractive (Nallapati et al., 2017) , abstractive (Nallapati et al., 2016; Zhang et al., 2019) and hybrid models (See et al., 2017; Gu et al., 2016) for summarization. Much of the recent work has shown these models to generate near-human coherent summaries while retaining reasonable factual correctness.

Dialogue summarization: While most neural summarization has focused on news corpora, recent work has tried to tackle unique challenges associated with summarizing dialogues. (Goo and Chen, 2018) proposes using dialogue history encoders based on the type of dialogue section to inform the generation. (Liu et al., 2019a) propose using key points as a means of categorizing sections of dialogue.

Medical dialogue summarization Existing work

(Alsentzer and Kim, 2018; Zhang et al., 2018; Liu et al., 2019b; Krishna et al., 2020a,b; Joshi et al., 2020) in this space focuses on effective summarization by incorporating medical knowledge from a modeling perspective. Our work also focuses on incorporating medical knowledge from a data labeling perspective. We show how we leverage pretrained language models and low-shot learning (Brown et al., 2020) to collect labeled data for medical dialogue summarization. We also show how this data can improve performance over models that are trained solely on existing human labeled data.

3 Background: Can GPT-3 serve as a medical summarizer?

Ignoring the privacy concerns and practitioner-in-the-loop considerations, we first explore whether GPT-3 (Brown et al., 2020) is a good medical summarizer by itself.

GPT-3 takes as input a *priming context* to perform the task on a previously unseen example. Priming context refers to the text description of a task and a few demonstrations of the task being accomplished (in our case, that would be dialogue snippet summarization).

Table 1 column 2 provides examples of summaries generated by the GPT-3 model. We can clearly see that it misses a number of important pieces of information in the snippets – first, missing medical concepts making the summary unusable (Rows 1-2). Second, the model may not always get the affirmations correct (Row 3). Third, the summary may repeat redundant information from the doctor's queries (Row 4).

Based on these observations, we might *prematurely* conclude that GPT-3 can not be used for medical summarization task. However, our key observation in exploring GPT-3 is that it is sensitive to the priming context (also reported in (Liu et al., 2021)), as the model does not learn but just adheres to the examples given. As we show in 4, we exploit this variability in GPT-3 output via ensembling and infusion of medical knowledge so that it can be used as a part of an effective low-shot learning approach to medical summarization.

4 Infusing Medical Knowledge in GPT-3 for use as a Data Generator

We are interested in a model that uses only a small amount of human labeled data to learn an effec-

Snippet	GPT-3	GPT-3-ENS
DR: Thank you so much for sharing. Are you on any regular medications for that? **PT**: My only regular medication is birth control - Apri. Low dosage.	On birth control.	Only regular medication is birth control -**Apri**. **Low dosage**.
DR: You had mentioned the headache starting after stopping the medications. Have you had similar headache while stopping medications before? **PT**: Yes **PT**: Well that's a complicated question **PT**: I'm not really sure **DR**: Okay, no worries. Wanted to know a bit more since you had mentioned about them. That's all.	Yes. Stopping medications before.	Yes, has **headache** while stopping medications before. **Is not sure.**
DR: Do you have pain when you notice penile discharge? **PT**: no i'm not	Did not notice penile discharge. No pain.	**Doesn't have pain when noticing penile discharge.**
DR: I have a few questions to ask. How long have you been having this vaginal discomfort? **PT**: only a few days, since like friday or saturday maybe	I have a few questions to ask. How long has she been having vaginal discomfort?	Has been having **vaginal discomfort for only a few days, since friday or saturday.**

Table 1: Input dialogue snippets along with summaries generated by GPT-3 in column 2 and our approach, GPT-3-ENS , in column 3.

tive medical dialogue summarizer. At the same time, we want such a model to be used in a practical practitioner-in-the-loop setting where medical correctness and patient privacy are of paramount importance.

In order to achieve these goals, we propose a two-pronged approach

1. Introduce **GPT-3-ENS** where we infuse medical knowledge into GPT-3 and use it within an inner loop to make it effective at medical summarization.

2. Leverage GPT-3-ENS as a data generator to obtain a large training set [2] to train an in-house medical dialogue summarization model. Such an in-house model can be used at inference

---

[2]Unlike data at inference time, training data is fixed and can be ensured to be privacy protected

time without the practical constraints related to protecting patient privacy that would require full de-identification to be applied in any conversation, if we were to access the GPT-3 service. It also lends itself well to the practioner-in-the-loop setting.

4.1 GPT-3-ENS : Medically-aware ensemble of GPT-3

As discussed in 3, GPT-3 is quite sensitive to the priming context. While one approach may be to provide GPT-3 with the most informative context for a task, this itself is a daunting task and can potentially be tackled if we had a large number of labeled examples (which is the exact problem we want to tackle with GPT-3).

Drawing on the learning from vast literature in ensembling techniques *c.f.* (Bishop et al., 1995), our first key insight is that if we can generate multiple summaries from GPT-3 using a variety of priming contexts, then we should be able to ensemble these outputs to identify the summary that is ideal for the dialogue. This insight leads to a question on how to ensemble multiple text summaries. The answer to this question relies on the core requirement for medical summarization: we care about the coverage of medical concepts mentioned and therefore the best ensembling function is the one that returns the summary with the most medical information in the dialog input.

In Algorithm 1 we provide our approach to the medically aware GPT-3 ensemble **GPT-3-ENS** . We assume access to a small set of labeled examples $\mathcal{L}$. For each input dialog snippet, D, we get K summaries, by invoking GPT-3 each time with N examples sampled randomly without replacement from $\mathcal{L}$. We also assume access to a medical entity extractor that can discern the medical concepts from both the dialogue snippet and the summary. The algorithm returns the best summary that has the highest recall in terms of capturing the medical concepts in the dialogue. For this purpose, we use an in-house medical concept extractor MEDICALEN-TITYRECOGNIZER that can identify medical concepts from a given piece of text. This extractor has access to the universe of medical concepts based on Unified Medical Knowledge Systems [3], which includes patient symptoms, disorders, laboratory tests and medications. Note that any medical entity recognizer (*cf.* (Fu et al., 2019) and references

---

therein) that has coverage for all these types of medical concepts found in medical conversations can be used.

Algorithm 1 Medically aware GPT-3 ensemble summarizer (**GPT-3-ENS**)

Require: dialogue snippet T, ensembling trials K, universe $\mathcal{L}$ of labeled examples, medical entity extractor $MedicalEntityRecognizer$, GPT3

1: $C^* \leftarrow MedicalEntityRecognizer(T)$
2: **for** $i \leftarrow 1, \cdots, K$ **do**
3:     $S \leftarrow$ sample N examples from $\mathcal{L}$
4:     $\texttt{summary}_i \leftarrow \texttt{GPT3}(S, T)$
5:     $C_i \leftarrow MedicalEntityRecognizer(\texttt{summary}_i)$
6: **end for**
7: $best \leftarrow \arg\max_i \frac{|C_i \cap C^*|}{|C^*|}$
8: **return** $\texttt{summary}_{best}$

Reconsider Table 1 for qualitative comparison between GPT-3 and GPT-3-ENS . We can see that summaries obtained using GPT-3-ENS capture the medical concepts comprehensively (shown in bold) and also have better grammatical structure. We also quantitatively validate the summaries on a small data set distinct from what is used for priming(see § 6.2 for guidelines). In Figure 2, based on doctor evaluation, we can see that GPT-3-ENS is significantly better at summarization than GPT-3 .

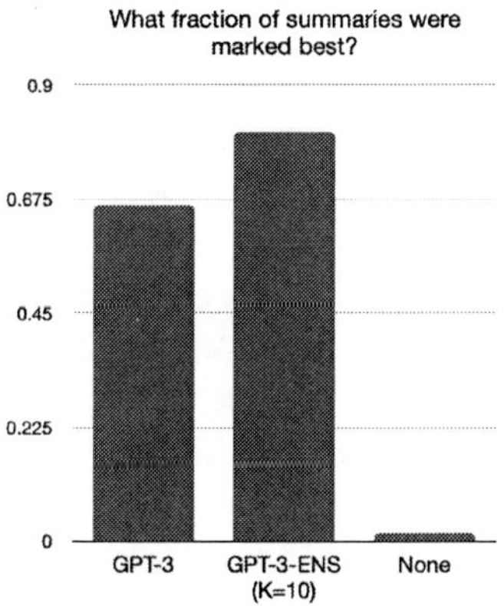

Figure 2: Doctor evaluation of which among GPT-3 and GPT-3-ENS summaries they considered "best" showing that GPT-3-ENS is a better approach for labeling

4.2 GPT-3-ENS as a data labeler

We use GPT-3-ENS described in 4.1 as our labeled data generator. In particular, we use our approach

---

[3]`https://www.nlm.nih.gov/research/umls/index.html`

to collect a large amount of labeled examples that serve as inputs to training an off-the-shelf summarization model. This resolves the concern of using GPT-3 in a real world application where the patient's conversation (in its raw form) needs to be exchanged with an external third party such as OpenAI/GPT-3 which may not have design/privacy regulations around HIPAA. In our approach, however, with the help of experts, it is easy to ensure that the dialogues that will used for priming as well as in the training set are chosen following privacy protocols.

5 Datasets

We collected a random subset of medical conversation dialogues from our chat-based telemedicine platform. Often medical conversation follows a linear ordering of medical history gathering (understanding patient symptoms) that enables creating the summary of the dialog by stitching together summaries of the snippets in chronological order (Joshi et al., 2020). Therefore, we split each dialogue into a series of local dialogue snippets using a simple heuristic: the turns between two subsequent questions by a physician corresponds to a snippet. The length of these snippets ranged anywhere from two turns (a physician question and patient response) to ten turns.

We had medical doctors[4] summarize these snippets. The doctors were asked to summarize the sections as they would for a typical clinical note by including all of the relevant history taking information. If a local snippet did not contain any history taking information it was excluded from annotations. For example in the beginning or end of conversations there may be turns that are purely greetings and not part of the patient history taking process. Further some snippets maybe purely educational in nature and are excluded as well. We eventually obtained a total of 6900 labeled snippet-summary pairs.

Human labeled dataset train/test split: From the 6900 labeled snippet-summary pairs (denoted as H_{6900}), we generated a randomly sampled test set $T = 500$ that we use in all our evaluations.

The dataset $H_{6900} - T$ is used to generate the priming dataset for GPT-3 related models as well as the datasets we use to train our summarization

models.

GPT-3-ENS dataset: Let $\mathbf{GCF}_p^k$ be the dataset of size p generated using GPT-3-ENS with k ensembling trials. To generate dataset $\mathrm{GCF}^{K=k}$, we require $\{\mathrm{H}_n\}_{i=1}^k$ datasets (note the independence on p), and thus $n \times k$ labeled examples for priming. These $n \times k$ examples are randomly sampled from the universe of human labeled examples $H_{6900} - T$. In our experiments, we sample without replacement so that no examples are reused across the k tries. To allow comparison between our experiments with different K values, we use the same seed for random sampling.

6 Evaluation Metrics

Multiple studies have shown that automated metrics in NLP do not always correlate well to human judgments as they may not fully capture coherent sentence structure and semantics (Stephen Roller, 2020; Kryściński et al., 2019). Since medical dialogue summarization would be used to assist health care, it is important for doctors to evaluate the quality of the output.

6.1 Automated metrics

While we measure model performance on standard metrics of ROUGE (Lin, 2004)[5], we also measure a model's effectiveness in capturing the medical concepts that are of importance, and their negations (Joshi et al., 2020)

Medical Concept Coverage: The concept coverage set of metrics captures the coverage of medical terms in the model's output summary with respect to the ground truth. In particular, let $\mathcal{C}$ be the set of medical concepts in the reference summary and $\hat{\mathcal{C}}$ be the set of concepts in the summary output by the model. Then, Concept recall $= \frac{\sum_{n=1}^{N} |\hat{\mathcal{C}}^{(n)} \cap \mathcal{C}^{(n)}|}{\sum_{n=1}^{N} |\mathcal{C}^{(n)}|}$ and Concept precision $= \frac{\sum_{n=1}^{N} |\hat{\mathcal{C}}^{(n)} \cap \mathcal{C}^{(n)}|}{\sum_{n=1}^{N} |\hat{\mathcal{C}}^{(n)}|}$.

We use these to compute a Concept F1[6] We use an in-house medical entity extractor to extract medical concepts in the summary. Medical concepts in the decoded summary that weren't present in the original conversation would be false positives and vice versa for false negatives.

---

[4]These are the same doctors who practice on the same telemedicine platform.

[5]We use the following package with default configuration:
`https://github.com/google-research/`
`google-research/tree/master/rouge`

[6]Note if there are no concepts detected in the snippet and summary by the entity extractor, then a conservative F1 score of 0 is given for that example.

Negation Correctness: To measure the effectiveness of the model to identify the negated status of medical concepts, we use Negex (Harkema et al., 2009) to determine negated concepts. Of the concepts present in the decoded summary, we evaluate precision and recall on whether the decoded negations were accurate for the decoded concepts and compute a negation F1[6].

6.2 Doctor Evaluation

We also had doctors, who serve patients on our telehealth platform, evaluate the summaries produced by the models. Given the local dialogue snippets and the generated summary, we asked them to evaluate the extent to which the summary captured factually correct and medically relevant information from the snippet. Depending on what percentage of the concepts were correctly mentioned in the decoded summary of the provided snippet, the doctors graded the summaries with *All* (100%), *Most* (at least 75%), *Some* (at least 1 fact but less than 75%), *None* (0%) labels.

We also formulated a comparison task where given summaries generated by different models and the associated dialogue, they were asked which summary was the "best" from a usability perspective. Usability was defined as whether the summary could stand in as a replacement for reading the dialogue snippet i.e. whether it captures the correct concepts from the snippet and whether the negations are accurate. The doctors had the ability to use "all" and "none" in this task depending on if all models being compared captured a good summary or if none of them did.

To avoid bias, the doctors do not know the model that produced the summary in both the experiments. In the comparison task, the summaries were provided in randomized order so that there is no bias in the order of presentation of the summaries.

7 Experiments and Results

Additional models considered: To evaluate the efficacy of **GPT-3-ENS** as a source of labeled data generator, we considered models with distinct objective functions for abstractive and hybrid (abstractive/extractive) summarization. We used **PEGASUS** (Zhang et al., 2019) for abstractive summarization and Dr. Summarize which we denote as **DRSUM** (Joshi et al., 2020) for extractive summarization. For **DRSUM** , we also use their best performing variant (referred as

2M-PGEN in (Joshi et al., 2020)) which penalizes generator loss and favors extractive copying.

Implementation Details: We used GPT-3 via the API released by OpenAI[7]. Maximum response length was set to 128 tokens, temperature to 0.6 and presence and frequency penalties both set to 0. For GPT-3-ENS , we use $K = 10$ ensembling trials for all our experiments, unless otherwise specified. We observed that $N = 21$ was the maximum number of examples we could prime GPT-3 with given the maximum context window length of 2048 tokens for the API. We therefore fix the size of our priming dataset to be 21 in all experiments which invoke GPT-3. Hence we set L to be a random subset of 210 examples from $H_{6900} - T$.

We followed parameter settings for DRSUM from (Joshi et al., 2020) for pretraining on the CNN-Dailymail dataset. We then fine-tuned on our summarization task dataset with a batch size of 16, source_max_tokens = 400, response_max_tokens = 200 and max_grad_norm clipped at 2.0, for two epochs with a learning rate of 0.15 using Adagrad optimizer.

We used the PEGASUS implementation that is pretrained on CNN-Dailymail[8] provided by (Wolf et al., 2020). We fine-tuned it on our summarization task dataset with an effective batch size of 256, source_max_tokens = 512, response_max_tokens = 128 for two epochs using Adafactor[9] optimizer at the default settings in Hugging Face. For both PEGASUS and DRSUM , we used a beam size of four for decoding.

7.1 Training summarization models using data labeled by GPT-3-ENS

We compare PEGASUS and DRSUM trained on human labeled data H_{6400} and GPT-3-ENS synthesized data $GCF_{6400}^{K=10}$. Note that synthesizing $GCF_{6400}^{K=10}$ needed all of $21 \cdot 10 = 210$ human labeled examples, where 21, as a reminder, is the maximum number of inputs that can be used for priming.

Table 2 compares quantitative performance of PEGASUS and DRSUM trained on these two datasets. The main observation is that with only

---

[7]https://beta.openai.com/

[8]https://huggingface.co/google/pegasus-cnn_dailymail

[9]https://huggingface.co/transformers/main_classes/optimizer_schedules.html#adafactor-pytorch

Models	Train Data Source	Metrics		
		Negation F1	Concept F1	ROUGE-L F1
PEGASUS	H_{6400}	21.09	35.96	55.59
	$GCF_{6400}^{k=10}$	**28.89**	40.02	53.43
	$GCF_{12800}^{k=10}$	26.70	40.21	56.66
	$GCF_{25600}^{k=10}$	28.61	**40.58**	**58.44**
DRSUM	H_{6400}	**26.75**	**39.95**	**52.70**
	$GCF_{6400}^{k=10}$	24.29	37.55	48.47
	$GCF_{12800}^{k=10}$	26.66	38.49	49.18
	$GCF_{25600}^{k=10}$	26.08	39.47	50.85

Table 2: Automated evaluation of summarization models trained with different data labeling methodologies. Note that the amount of human labeled data is still pretty low (210), compared to 6400 when we do not use our approach.

210 human labeled examples, our approach GPT-3-ENS is able to generate a large amount of training data for both pre-trained summarization models, PEGASUS and DRSUM , in such a manner that they yield comparable (or better perfomance) than if they had been trained with only $6400(\sim30x)$ human labeled examples.

For PEGASUS , the summarization performance improves drastically compared to model fine-tuned using only the human labeled data. We hypothesize that data generated from GPT-3-ENS can serve as quality training data for abstractive models such as PEGASUS but not so much for hybrid models such as DRSUM due to GPT-3 being a generative language model. The summaries written by our human doctors have writing structure similar to that of a hybrid summarization model such as DR-SUM that is more extractive in nature. This can explain why DRSUM did not show performance gain when using generated data from GPT-3-ENS . The key, however, is that it still did perform *on par*.

In the same Table 2, we also present the results with increased amounts of data (12800 and 25600) from GPT-3-ENS . There is little or no further improvement in the automated metrics of concept and negation F1. However, ROUGE-L F1 improves reflecting the improvements in coherency of the summaries. We leave this area as future work to explore.

7.2 Effect of combining human labeled data with data labeled by GPT-3-ENS

Since GPT-3 relies on limited local priming context ($N = 21$) it may not be agile in providing robust summaries for a multitude of variations in snippets, focusing on the exploitation part of the exploration-exploitation trade-off. We hypothesize that best summaries then will be synthesized by a model trained on a dataset with human and GPT-3-ENS labeled examples. To evaluate this, we introduced a mixing parameter α, the ratio of GPT-3-ENS labeled examples to human labeled examples. For instance, with 6400 human labeled examples, $\alpha = 0.5$ implies the dataset contains 6400 human labeled examples along with $0.5 * 6400 = 3200$ GPT-3-ENS generated examples. We experiment with $\alpha = 0.5, 1, 2, 3$.

From Table 4, we observe that for both PEGA-SUS and DRSUM , mixture of human labeled and GPT-3-ENS data consistently improves almost all automated metrics for all α values[10] The lift in metrics is lower for DRSUM , again illustrating the idea we highlighted in § 7.1 of GPT-3-ENS data being more amenable to abstractive models such as PEGASUS than for hybrid or extractive-biased models such as DRSUM . Table 3 provides qualitative comparison between summaries generated by each of these models.

For simplicity, we chose the smallest GPT-3-ENS mix i.e. $\alpha = 0.5$ for human evaluation where we ask doctors to evaluate summaries from model trained on human, GPT-3-ENS and human+GPT-3-ENS data. Figure 3 and Figure 4 show that doctors prefer summaries from the model trained on the mixture data over those produced by models trained on human or GPT-3-ENS data alone, in terms of amount of medical information captured as well as the overall quality of the summary. Furthermore, Figure 3(b) also shows that for PEGASUS , doctors prefer the summaries from a model trained on $GCF_{6400}^{K=10}$ (which needed *only 210 human labeled examples*) over those produced by a model trained on 6400 human labeled examples.

8 Conclusion

We introduced a medically-aware GPT-3 data labeler, GPT-3-ENS , for the task of medical conversation summarization. At the heart of the approach is a medically aware ensembling criterion that ensembles multiple summaries for an input from a powerful low-shot learner such as GPT-3. We showed that this approach can generate quality

---

[10]Note here that the claim is not that increasing α improves metrics but that mixing GPT-3-ENS and human labeled data improves metrics over models trained only using human data. We leave it as a future work on how to trade-off between human and GPT-3-ENS labeled data.

Snippet	Model trained on H_{6400}	Model trained on $GCF_{6400}^{K=10}$	Model trained on H_{6400}+$GCF_{3200}^{K=10}$
DR: Have you ever been tested for any underlying health conditions such as diabetes, hypothyroidism or polycystic ovarian syndrome? **PT**: No **PT**: I have been told I have prediabetes	Has not been tested for any underlying health conditions.	Hasn't tested for any underlying health conditions such as diabetes, hypothyroidism or polycystic ovarian syndrome	Has not been tested for any underlying health conditions. Has been told has prediabetes.
DR: DR: Do you have pus appearing discharge from the site? **PT**: Yes. If the bubbles pop it leaks out a watery substance	Has pus appearing from the site.	Pus appearing from the site	Pus discharge from the site. If bubbles pop it leaks out a substance.

Table 3: Input conversation snippets along with summaries generated by models trained on different data

Models	Train Data Source	Metrics		
		Negation F1	Concept F1	ROUGE-L F1
PEGASUS	H_{6400}	21.09	35.96	55.59
$\alpha = 0.5$	$H_{6400} + GCF_{3200}^{K=10}$	30.14	43.49	**62.45**
$\alpha = 1$	$H_{6400} + GCF_{6400}^{K=10}$	30.70	43.73	60.63
$\alpha = 2$	$H_{6400} + GCF_{12800}^{K=10}$	29.43	41.02	59.85
$\alpha = 3$	$H_{6400} + GCF_{25600}^{K=10}$	**31.93**	**44.68**	61.05
DRSUM	H_{6400}	26.75	39.95	52.70
$\alpha = 0.5$	$H_{6400} + GCF_{3200}^{K=10}$	**27.51**	40.46	**53.39**
$\alpha = 1$	$H_{6400} + GCF_{6400}^{K=10}$	27.18	40.36	51.00
$\alpha = 2$	$H_{6400} + GCF_{12800}^{K=10}$	27.19	**40.68**	53.07
$\alpha = 3$	$H_{6400} + GCF_{25600}^{K=10}$	26.33	39.89	52.29

Table 4: Combining human labeled datasets with datasets generated using our proposed approach

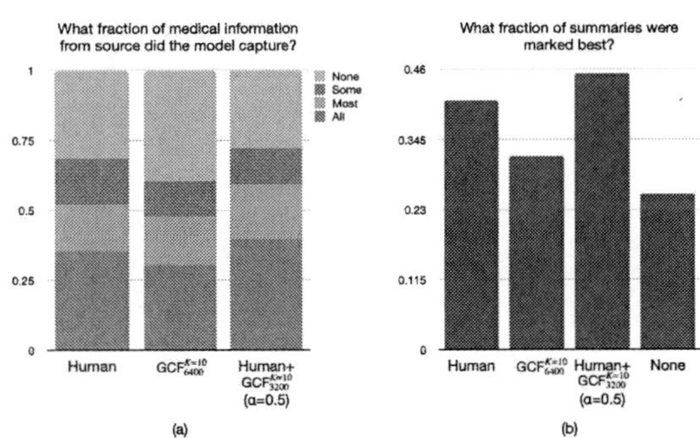

Figure 4: Doctor evaluation of amount of medical information covered by summaries provided by DR-SUM models and which ones they considered "best"

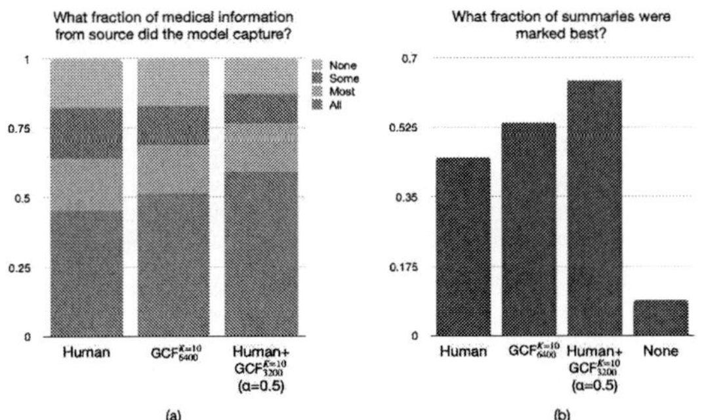

Figure 3: Doctor evaluation of amount of medical information covered by summaries provided by PEGASUS models and which ones they considered "best"

training data for medical dialogue summarization models while ensuring medical correctness. We show that using a very small number of human labeled examples, 210, we are able to produce more medically correct and better quality summaries than using roughly thirty times as many human labeled examples for two different summarization models. In this work we used a simple ensembling technique that dialogue summaries should retain all the medical information discussed in the dialogue. Future work could be to improve our ensembling function to take into account other medical priors such as affirmations and importance/relevance of the information in the dialog.

References

Emily Alsentzer and Anne Kim. 2018. Extractive summarization of EHR discharge notes. *CoRR*, abs/1810.12085.

Christopher M Bishop et al. 1995. *Neural networks for pattern recognition*. Oxford university press.

Tom B Brown, Benjamin Mann, Nick Ryder, Melanie Subbiah, Jared Kaplan, Prafulla Dhariwal, Arvind Neelakantan, Pranav Shyam, Girish Sastry, Amanda Askell, et al. 2020. Language models are few-shot learners. *arXiv preprint arXiv:2005.14165*.

Sunyang Fu, David Chen, Sijia Liu, Sungrim Moon, Kevin J. Peterson, Feichen Shen, Yanshan Wang, Liwei Wang, Andrew Wen, Yiqing Zhao, Sunghwan Sohn, and Hongfang Liu. 2019. Development of clinical concept extraction applications: A methodology review. *CoRR*, abs/1910.11377.

Chih-Wen Goo and Yun-Nung Chen. 2018. Abstractive dialogue summarization with sentence-gated modeling optimized by dialogue acts. *CoRR*, abs/1809.05715.

Jiatao Gu, Zhengdong Lu, Hang Li, and Victor O.K. Li. 2016. Incorporating copying mechanism in sequence-to-sequence learning. In *Proceedings of the 54th Annual Meeting of the Association for Computational Linguistics (Volume 1: Long Papers)*, pages 1631–1640, Berlin, Germany. Association for Computational Linguistics.

Henk Harkema, John N. Dowling, Tyler Thornblade, and Wendy W. Chapman. 2009. Context: An algorithm for determining negation, experiencer, and temporal status from clinical reports. *Journal of Biomedical Informatics*, 42(5):839 – 851. Biomedical Natural Language Processing.

Anirudh Joshi, Namit Katariya, Xavier Amatriain, and Anitha Kannan. 2020. Dr. summarize: Global summarization of medical dialogue by exploiting local structures. *arXiv preprint arXiv:2009.08666*.

Kundan Krishna, Sopan Khosla, Jeffrey P. Bigham, and Zachary C. Lipton. 2020a. Generating soap notes from doctor-patient conversations.

Kundan Krishna, Amy Pavel, Benjamin Schloss, Jeffrey P Bigham, and Zachary C Lipton. 2020b. Extracting structured data from physician-patient conversations by predicting noteworthy utterances. In *Explainable AI in Healthcare and Medicine*, pages 155–169. Springer.

Wojciech Kryściński, Bryan McCann, Caiming Xiong, and Richard Socher. 2019. Evaluating the factual consistency of abstractive text summarization.

Chin-Yew Lin. 2004. ROUGE: A package for automatic evaluation of summaries. In *Text Summarization Branches Out*, pages 74–81, Barcelona, Spain. Association for Computational Linguistics.

Chunyi Liu, Peng Wang, Jiang Xu, Zang Li, and Jieping Ye. 2019a. Automatic dialogue summary generation for customer service. In *Proceedings of the 25th ACM SIGKDD International Conference on Knowledge Discovery Data Mining*, KDD '19, page 1957–1965, New York, NY, USA. Association for Computing Machinery.

Jiachang Liu, Dinghan Shen, Yizhe Zhang, Bill Dolan, Lawrence Carin, and Weizhu Chen. 2021. What makes good in-context examples for gpt-3?

Zhengyuan Liu, Angela Ng, Sheldon Lee, Ai Ti Aw, and Nancy F. Chen. 2019b. Topic-aware pointer-generator networks for summarizing spoken conversations.

Devin M Mann, Ji Chen, Rumi Chunara, Paul A Testa, and Oded Nov. 2020. COVID-19 transforms health care through telemedicine: evidence from the field. *Journal of the American Medical Informatics Association*. Ocaa072.

Ramesh Nallapati, Feifei Zhai, and Bowen Zhou. 2017. Summarunner: A recurrent neural network based sequence model for extractive summarization of documents. In *Proceedings of the Thirty-First AAAI Conference on Artificial Intelligence*, AAAI'17, page 3075–3081. AAAI Press.

Ramesh Nallapati, Bowen Zhou, Cicero dos Santos, Çağlar Gu̇̀‡lçehre, and Bing Xiang. 2016. Abstractive text summarization using sequence-to-sequence RNNs and beyond. In *Proceedings of The 20th SIGNLL Conference on Computational Natural Language Learning*, pages 280–290, Berlin, Germany. Association for Computational Linguistics.

Weizhen Qi, Yu Yan, Yeyun Gong, Dayiheng Liu, Nan Duan, Jiusheng Chen, Ruofei Zhang, and Ming Zhou. 2020. Prophetnet: Predicting future n-gram for sequence-to-sequence pre-training.

Abigail See, Peter Liu, and Christopher Manning. 2017. Get to the point: Summarization with pointer-generator networks. In *Association for Computational Linguistics*.

Tait D. Shanafelt, Lotte N.Dyrbye, Christine Sinsky, Omar Hasan, Daniel Satele, Jeff Sloan, and Colin P. West. 2016. Relationship between clerical burden and characteristics of the electronic environment with physician burnout and professional satisfaction. *Mayo Clinic Proceedings*, 91:836–848.

Naman Goyal Da Ju Mary Williamson Yinhan Liu Jing Xu Myle Ott Kurt Shuster Eric M. Smith Y-Lan Boureau Jason Weston Stephen Roller, Emily Dinan. 2020. Recipes for building an open-domain chatbot.

Ilya Sutskever, Oriol Vinyals, and Quoc V. Le. 2014. Sequence to sequence learning with neural networks. In *Proceedings of the 27th International Conference on Neural Information Processing Systems - Volume 2*, NIPS'14, page 3104–3112, Cambridge, MA, USA. MIT Press.

Thomas Wolf, Lysandre Debut, Victor Sanh, Julien Chaumond, Clement Delangue, Anthony Moi, Pierric Cistac, Tim Rault, Rémi Louf, Morgan Funtowicz, Joe Davison, Sam Shleifer, Patrick von Platen,

Clara Ma, Yacine Jernite, Julien Plu, Canwen Xu, Teven Le Scao, Sylvain Gugger, Mariama Drame, Quentin Lhoest, and Alexander M. Rush. 2020. Transformers: State-of-the-art natural language processing. In *Proceedings of the 2020 Conference on Empirical Methods in Natural Language Processing: System Demonstrations*, pages 38–45, Online. Association for Computational Linguistics.

Jingqing Zhang, Yao Zhao, Mohammad Saleh, and Peter J Liu. 2019. Pegasus: Pre-training with extracted gap-sentences for abstractive summarization. *arXiv preprint arXiv:1912.08777*.

Yuhao Zhang, Daisy Yi Ding, Tianpei Qian, Christopher D. Manning, and Curtis P. Langlotz. 2018. Learning to summarize radiology findings. *CoRR*, abs/1809.04698.

Snippet	Summary	Prompt
PT: Today spit out a bit of mucus and noticed a bit of blood. **DR:** Okay, how long have you been on these medications? **PT:** About 2 years	Has been on these medications for about 2 years	Today spit out a bit of mucus and noticed a bit of blood.[STOP] Okay, how long have you been on these medications?[SEP]About 2 years[SUMMARIZED]Has been on these medications for about 2 years.[STOP]
DR: Is the bleeding from the anal opening and not the vagina? Has something similar happened before? **PT:** yes from the anal opening	The bleeding is from the anal opening.	Is the bleeding from the anal opening and not the vagina? Has something similar happened before?[SEP]yes from the anal opening[SUMMARIZED]The bleeding is from the anal opening.[STOP]

Table 5: Prompt for GPT-3 given two examples

A GPT-3 Prompt

We utilize a fairly simple prompt to have GPT-3 generate summaries. Each example (snippet_text, summary_text) is concatenated to the empty string with the following transformation: "{snippet_text}[SUMMARY]{summary_text}[STOP]" to form the prompt. We seperate the conversational turns in snippet_text with the "[SEP]" token. Table 5 shows a prompt that would be generated and used to prime GPT-3 given two examples. As mentioned in § 7 in our experiments we use 21 examples to generate a prompt

Author Index